Quantum Energetics and Spirituality

Quantum Energetics and Spirituality

Aligning with Universal Consciousness

Volume 2

KENNETH SCHMITT

Copyright © 2021 by Kenneth Schmitt

All rights reserved. No part of this book may be reproduced or transmitted in any form or by any means, electronic or mechanical, including photocopying, recording, or by any information storage and retrieval system, without permission in writing from the author.

ISBN: 9798985106404

Kenneth Schmitt
Phone: +1-808-280-4041
Email: timeless1@twc.com
Website: https://www.ConsciousExpansion.org

Contents

Introduction   1

1. The Power of Our Perspective   3
2. Moving into Ascension   53
3. Working with the Law of Attraction   91
4. Expanding Consciousness   101
5. Self-Realization   135
6. Universal Consciousness and the Quantum Field   173
7. Aligning with Higher Consciousness   209

Introduction

In our culture there has been a great divide between science and spirituality. Scientific knowledge has been limited to the empirical world, and the human mind has been thought to be an operation of the physical brain, at least until the birth of quantum physics at the turn of the twentieth century. With the simple double slit experiment with photons, scientists discovered the quantum field of all potentialities and universal consciousness. Physics became mysterious and ultimately challenging to the established scientific community. Although they must accept its proof, most physicists today refuse to recognize its implications for our understanding of reality, because doing so would mean that the empirical world is an electromagnetic field of quantum energy held in place by the conscious participation of all members of humanity. The leap in understanding is too great.

This knowledge had been the exclusive purview of Christian mystics, Islamic Sufis, Tibetan Mahayana Buddhist Rinpoches, yogis, shamans and spiritual adventurers. Now, however, science and spirituality have united in understanding that consciousness is the basis of everything. The human brain is only an instrument of our consciousness, which is beyond the physical. From a practiced position of emptiness or zero-point consciousness, we can create anything that we imagine and feel, in complete alignment with its vibratory patterns. By gaining absolute control of their attention, spiritual masters can conjure up the physical presence of deities and ascended Beings.

The world we experience is a creation of our own being, arising from the unified quantum field through our recognition of energetic wave patterns that we imagine and feel. These wave patterns appear physical through our recognition and interpretation within our own consciousness. They appear and disappear innumerable times every second and seem solid, just as individual images slide through a movie projector and appear to be a flow of moving scenes.

What is real is consciousness. It is all-pervasive throughout the cosmos and perhaps beyond. Universal consciousness expresses itself as the unified quantum field of all potentialities. This is a plasma electromagnetic expanse of wave patterns, radiant with photons and life force. The seers and mystics perceive that there is a quality of unconditional love that is inherent in universal consciousness. This has not yet been recognized by quantum physicists, but it is implied by the organization of the living cosmos. Everything is conscious and works in perfect harmony, with the exception of humanity, due to our desire to use our free will to explore anomalous vibrations under the guise of self-created limitations to our awareness.

We inhabit a living planet with a resonant frequency that has remained historically at 7.83 cycles per second, but this is now increasing under the influx of massive waves of gamma ray photons and solar ejections that are raising the resonant frequency far beyond its historical pattern. In order to continue living here, humanity must raise its resonant frequency in alignment with the Earth. This brings a change in consciousness to higher vibrations of peace, joy and compassion. It awakens an awareness of expanded consciousness and brilliant Self-identity. We are moving beyond the empirical illusion into a magical realm of ethereal beauty and love.

This book is a guide to making this reality adjustment in our lives.

1.

The Power of Our Perspective

Personal and Collective Opportunity

From quantum physics experiments, we know that sub-atomic particles can be in more than one place at the same time. This is possible in the quantum field, but not if human consciousness or its technological extension is observing. We have programmed ourselves to interpret a limited spectrum of energy in the quantum field as empirical reality. Since sub-atomic entities can manifest either as patterns of electromagnetic waves or as particles, by implication everything that they comprise has the same quality.

Are we actually a part of empirical reality, or are we just observers of this entire spectrum of vibrations without limitations? We may just be playing parts in a drama that we're involved in emotionally and mentally. It seems like the entirety of our reality. We have been masterful in limiting our conscious spectrum of vibrations in order for us to have experiences that we could never have, if we actually knew who we are beyond the empirical spectrum. But now it is time to end the historical spectrum of resonance and depart from the drama of the world we have known.

The Shumann resonance of the Earth is rising to a higher spectrum of vibrations. This requires a leap in awareness on the part of humanity in order to remain conscious on this planet. We are intentionally becoming aware of a higher spectrum of emotions. These are all emotions without fear. They all inherently feel attractive to us, and they all vibrate with the frequencies of gratitude, love and joy.

If we can elevate the frequencies of our energy signatures to vibrate in resonance with gratitude, love and joy, we can actually live in this spectrum, as it elevates our experiences magically. This is possible because everything about us arises and demises in cycles from the quantum field continuously. We modulate all of the energies in our awareness. That's the reason that energy waves in a certain spectrum change to particles in our perception when we recognize them as such. By our very nature, we change the energy that we recognize all the time. Our predominant vibrations of thought and emotion are constantly creating for us new experiences of the same quality as the awareness of the creator. By intentionally imagining high-vibration scenarios and feeling ourselves living in them, the quantum field manifests the changed energy spectrum. In this way we create our new lives.

Once we know our true identity, we have a much more complete perspective on life. We have the conscious choice of living a beautiful and fulfilling life and being able to share this energy with everyone we encounter.

Adjusting Our Perspective

How do we gain the most beneficial perspective for ourselves?

Chapter 1. The Power of Our Perspective

Let's look at the nature of a perspective. It arises from our understanding of ourselves, of who we believe we are and what we believe are our capabilities.

In our current matrix-dreamed reality we believe we are completely limited in our consciousness to awareness of a limited spectrum of energies that our consciousness interprets as empirical. These are the vibrations that stimulate all of our senses, which are intimately connected to our emotional nature. Our ego-mind consciousness cannot easily allow our awareness to penetrate beyond our beliefs about ourselves; however, our emotional nature is not so tightly bound to the ego, and through our feelings, we can transcend our current experience. We do have emotional boundaries that were formed by inherited and experiential serious difficulties. We need to resolve these by transforming their energies into feelings of goodness and joy. These high-vibration feelings will carry our imaginations into visions of magic and wonder.

The transformation process begins by recognizing our boundaries, both emotional and our beliefs about ourselves. We need to question seriously all of them. If possible, we may realize their origins and purpose, and ask if we want to continue to feel anything that came from fear. In order to transform the fear of pain and mortality, we need to know that we are eternal beings. This is a leap in consciousness.

In our emotional being, we can just keep getting higher by intentional seeking of the fullness of knowing unconditional love. In this state we can recognize that our conscious awareness is unlimited. We can experience anything in the quantum field that we can feel and imagine. We can continue to move toward higher-frequency feelings and experience wonderful lives.

Who do we find that we are? We are pure awareness. We express ourselves as patterns of electromagnetic waves in the quantum field. Each of us has an energy signature that vibrates within a spectrum of frequency. Out of this energy we express our physical presence, which we recognize as material. The empirical world can stimulate our emotions, and our emotions also create empirical experiences. By using our imagination together with our emotions, we become modulators of any energies we choose to be aware of. We can enhance their intensity by aligning with the energy patterns, or we can change the vibratory patterns of any energy that we choose to come into alignment with. We can intentionally choose high-vibrational emotions. From this perspective, we may choose to feel being in our current situations, while also being aware of knowing universal consciousness in the quantum field. We can know that everything has consciousness. Everything arises continuously from the One Being, whose consciousness expresses itself as all the energy patterns in existence and who is present with life-giving intent in awareness of the experiences of every conscious entity, including us. We are the personalized presence of our Creator.

Transforming Our Relationships

We have relationships with everyone and everything we recognize. What is the nature of a relationship? It is an exchange of energy vibrations by the beings who recognize each other. We live within an energy spectrum or dimension that is held in conscious awareness by all of humanity. Every human on Earth participates in this energy and interacts with its wave patterns and amplitudes.

As we become aware of higher frequency emotions and let them draw us into greater joy, we attract relationships of trust, kindness and gratitude. We also radiate this energy from our energy signatures. As more and more of us open our awareness to greater love and joy, we transform the consciousness of humanity, as we all rise into a higher spectrum of greater joy. What is most important is that we feel high-frequency emotions and recognize them. This is the key to greater awareness, and this is where our perspective is going. Although we influence, through our energy signature, everything we recognize, we can have an even greater impact when we want to create something specific, through an intentional projection of energy, in which we use our creative imagination and emotions.

Every form of electromagnetic energy exists in the quantum field of all potentialities. As we observe energy patterns, they become empirical for us. When we stop recognizing them, they return to being energy patterns in the quantum field.

By recognizing these energies, we draw them into our experiences. If we are willing to realize unconditional love in our being, we become aware that we are infinitely powerful creators of whatever we can imagine and feel good about. We fill our lives with beauty and abundance, which we share with all. Constantly we modulate the energies around us with our energy signature and our intentions and feelings. Our intuition is clear for any interaction which is loving and joyous.

In every area of our lives, we can live in deepest love, gratitude and joy, including our relationships with all of the things in our lives and our possessions, including money. Money is currency—it's designed to flow. We enhance everything we recognize with the vibration of our perspective. This is how our attention is creative in modulating the energy patterns we recognize.

The Danger of Being Against Evil

Quantum physics has proved that electromagnetic wave patterns become material when we observe them. If we don't give them our attention, they continue as wave patterns in the quantum field of all potentialities. To this extent we are creators. Our recognition brings into our empirical experience anything we can intend to see. This occurs in every spectrum of vibration where we are consciously aware. Everything out to infinity is available to us. The quantum field contains all potentialities—everything that could possibly happen anywhere. We bring into our reality what we recognize.

Many movements around the world have been uprisings against mistreatment and enslavement. What happens with the energy of uprisings? It always works the same way. Those in control of the population trick the people into allowing themselves to be controlled. Then they become parasites until there is nothing left of the victims. Along the way the victims occasionally rebel. The heightened energy of the life force of the victims is creating more powerful energy for the focus of their attention--the energy of the oppressor, because the people recognize the reality of the oppressor and give it their life force through their anger and fear.

There is nothing keeping the victims from disappearing from the parasites vibrationally. It is only their internal programming that keeps them enthralled. What can the victims do to escape? They can stop recognizing the oppressor and withhold their life force. The oppressor loses energy and disappears. The people can imagine themselves in a different world, a world of light and love and joy. If they can hold this focus, they no longer are attractive to the lower-vibrational beings. The magnetism has shifted and the energy signatures of the victims have risen in vibration into freedom. The visions of the people are their recognition of the component energies in the quantum field, which brings the vibration of the visions into materialization in their experience.

Personal Quantum Energetics

On the path to a higher vibration of living, we occasionally need to pause and just Self-realize. We can release all of our worldly concerns and focus entirely within our Being. We can use our imagination to create the best scenarios we want to experience and the most intense feelings of gratitude and joy. Even if we do this only momentarily whenever we can bring ourselves to do it, we find ourselves doing it easier and better more and more. We can imagine ourselves beyond our current boundaries and in this way transform them into more expansive vibrations of higher frequencies in resonance with the energy of our heart. We are creating a new reality of joy and wonder. This is the energy that humanity is moving into and is being assisted by the increasingly more intense incoming gamma ray very-high-frequency photons that are greatly increasing the resonating frequency of the Earth.

Fighting in support or against low-frequency energies is pointless. We cannot involve ourselves in low-frequency patterns of energy and also stay in a high-frequency spectrum of vibrations. Some of us may choose to do so initially in an attempt to raise the resonant vibrations of the destructive energies or terminate them. The preferred way of raising the resonant frequency of any situation is to confront it as our higher Self, the divine One. By being present in high-frequency emotional energy, we transform the lower-frequency patterns. They either align with the higher frequencies, or they become unstable and dissolve. We can achieve a state of equilibrium within by resolving all dissonant energies that arise and by modulating them into alignment with higher consciousness or dissolving them. We are seeing this begin to happen in society. This is how energetics work in the quantum field and then manifest empirically.

The most important intention for us now is to Self-realize—to know and feel who we are beyond our bodies, and to become aware of the quality of the highest vibrations in our energy signature, what it feels like to be our higher Self. We realize that we are eternal beings, expressing ourselves as our personal energy signature, a portion of which is manifesting as our physical presence living in the spectrum of humanity's energy signature. We can realize our sovereignty. We cannot be threatened. We are eternal, absolute Beings, who are complete in ourselves in every way. We can create anything that we feel we need or want, always for the good of all participants. In universal consciousness, that means all conscious beings. We are unlimited in the empirical world, because we know how energetics work. It's what we all know intuitively. By aligning ourselves with our high-frequency emotions and imaginings, we attract energy patterns that resonate with us. As we keep raising the frequency of our energy signatures, we live more high-frequency joyful and loving lives.

Life is not a sequence of random events. We attract every event in our lives by the vibrations of our constantly-changing energy signature. As more of us vibrate at higher frequencies, it significantly raises the vibrational level of humanity's energy signature and stimulates awakening.

Personal Fulfillment

We all want a better life. Each of us may imagine it differently, but we all want our physical needs taken care of. We want to share our lives with the ones we love the most and want to be present with. We want health and well-being and ultimately everlasting life. Above all we want to preserve and protect our consciousness. We want to enjoy life forever in greatest love and joy. If we were designed to have these experiences, let's get with it! Every human can be completely fulfilled forever.

Achieving fulfillment depends upon our state of being in our own consciousness. We express ourselves as quantum energetic configurations represented by our predominant thoughts, feelings, beliefs and perspectives. Our energetic signature radiates our vibrations into the quantum field and exchanges wave patterns with other conscious entities. We attract energy patterns that resonate with our own.

If we want to elevate the quality of our lives, we need to elevate the frequency of our energy signature. Our energy generator is our emotional body, which identifies all of the energies around us and also expresses the energy that we radiate. It is our creative power source, and it issues from our heart. Through compassionate wisdom we must clear it of all of its scars and traumas accumulated in our lower vibrational experiences. We've learned many lessons and don't need to repeat them.

Now we can adjust our perspective on life. Our quest takes us into higher and higher vibrations of feelings and visions, seeking to expand our awareness into greater joy and love. This is the spectrum of high-frequency vibrations that leads to our fulfillment in every way. It is the dimension of knowing our true Being beyond our embodiment in time and space. We have a multi-dimensional presence and can be aware of the realm of beauty, joy, gratitude and bliss, while also being present in our current world, but with a transformed perspective.

This process is open to all of us, and we'll all be drawn this way eventually. Ultimately each of us travels this path within our own being. We become aware of our ability to transform low-frequency energy patterns that come into our attention, if we can maintain a high-frequency perspective. Our visions and emotions can align with a beautiful energy spectrum of brilliant feelings and experiences. We become attractive to resonant high-frequency wave patterns of joy and abundance, because we are flowing in alignment with the energy of the Creator, expressing itself in the quantum field as the life-giving consciousness of unconditional love. This is the essence of our Being. When we know and experience this level of frequency, everything we command and love comes to us.

How Are We All One Being?

We have all participated in designing and implementing the world that humanity experiences. We have been embodied many times and have played many roles, from the most despicable scoundrels to the brightest shining stars. We have played with dark energies of the worst kind and have been tortured for our strongest desires for freedom and higher consciousness. Now it's time for us to forgive ourselves and everyone else. From the depths of our being we can be in compassion and forgiveness.

The empirical world is so convincing as reality that we really believe we are separate beings. In the greater reality, however, we all arise from the same Being and share our essence with our Creator. In our consciousness and life force, we are all reflections of each other on an energetic level.

What is important is that we don't get stuck in low-frequency vibrations. Some beings have sought exorbitant amounts of money and power over others and have greatly mistreated those that they have conquered. This is where we've all been enslaved. For the victims to be angry, frustrated, hateful, depressed and judgmental keeps them in low vibrational experiences. Holding grudges is being stuck in low-vibration experiences. Forgiveness is necessary for ourselves for misguiding us by taking us into dark experiences that have created strong low-vibration emotions in us. These emotions can be transformed by recognizing them from a high-vibration perspective of compassionate wisdom. Our strong high-vibration presence will require low-vibrations to come into higher alignment, when encountering each other, or the low vibrations will destabilize and dissolve.

We do not condone low-vibrational actors and their energy. We can withdraw our attention and our life force from them. The way out of poverty, victimization and oppression is by raising our vibrations through imagining and feeling forgiveness and love and even joyful scenarios for ourselves. These vibrations radiate throughout our being and reflect back to us as higher-vibrational experiences in miraculous ways. By living in our high-vibration inner world, we create a high-vibration world of experience in our conscious awareness. This becomes our reality.

Chapter 1. The Power of Our Perspective

Analyzing Our Life Path

We are always free to choose our next path in life, and our intuition will always guide us in each moment to make the best moves along that path, until we decide to change it. Our only obstacles are our own attachments and beliefs about ourselves. If we intend to follow our inner guidance most beneficially, it is imperative to become clear within our own being.

Because we are energetic beings with our own personal energy signature, we vibrate at our own frequency. We attract other beings and circumstances that vibrate in resonance with us. We all have free choice always, in every circumstance. When we choose to express ourselves according to our highest passions, we are always guided intuitively toward the energetic manifestations that resonate with our heart-felt intentions.

If we identify with our ego's desires, preferences and personal beliefs, we cannot be open to our divine guidance. If we identify with our current circumstances, we continue to create more of the same. We have all been stuck in fear of some kind and rejection by others. By choosing intentionally or passively to subject ourselves to persons and circumstances that do not synchronize with us, we keep ourselves from personal fulfillment. As long as we feed our ego feelings of inferiority and need for someone or something outside of ourselves, we cannot recognize our true Being.

The matrix that we have chosen to live within is designed to keep us entranced in Humanity's low-vibration situations of injustice, poverty and enslavement. If we desire to live in a world of higher vibrations, a world of peace, love and joy, we must align our own being with these vibrations. All of life operates energetically. Our presence in our physical consciousness is only limited by our self-imposed limitations and boundaries. If we choose to move beyond all of this and into unlimited awareness and unconditional love, we must set our intention and heart-felt desire for this experience, and then open ourselves to our intuitive guidance without any distractions or attachments to persona, places and things. Once we begin to follow our inner guidance, our lives can become miraculously fulfilling in every way, regardless of our current circumstances.

Understanding the World We Inhabit

According to quantum physics, the world of our empirical senses is a complex structure of energy patterns that we all recognize and interpret as the reality of human experience. This conscious recognition could be no more substantial in its essence than a hologram. All of our senses are being stimulated by the vibrations that are within their operating energetic patterns. We have cooperatively created the forms and qualities of all of it by the operations of our consciousness. Everything we observe is a projection of our own consciousness. Without our conscious interpretation of wave patterns and vibrations, there would just be energy that we don't recognize. Then where would we be? We would not be aware of anything, except our awareness. This is who we are. If we ever get to this place, it will change all of our false beliefs about ourselves.

We're not finished evolving and expanding by knowing our true identity within the limits of our ego awareness. There is more. We exist with our personal energy signature in the unified quantum field. How do we find our way around, when we don't know anything except who we are? We use our imagination and our emotions. We create the quality of our life and let the quantum field fill in the resonant experiences. We can also create forms and scenarios. Everything we see on TV and in the movies is a creation of someone's imagination and emotional involvement. We all do this all the time, without knowing what we're doing. Humans believe that everything happens haphazardly, and that we're vulnerable to tragedy. And so it is, only it is not true. It's what is believed that creates the recognition and interpretation.

What would life be like without beliefs and interpretations about ourselves? Would we be anybody? We might not have a body. We can, if we want one. We can create it in any form or kind of being that we want, and we will be attracted to the realm where others are similar. We will be living in the field of unlimited potentiality, in which we can experience anything our heart imagines.

Being in the World, but Not of It

Once we realize that we are living in an artificial matrix of energies that we constantly modulate with our own recognition, we can open our awareness to realms beyond. We no longer need our limited beliefs about ourselves and the nature of our reality. We no longer need to feel small and vulnerable. We can imagine more possibilities for our conscious awareness. We can create scenarios that we recognize and feel as real. With practice these levels of vibrations radiate out into the quantum field for manifestation of the qualities of experience that we feel and imagine.

There are an infinite number of life experiences available to us in every moment. We get to live with those experiences that resonate with our conscious state of being, how we recognize ourselves to be. We express this in our personal energy signature. All of our beliefs about ourselves are self-imposed in order to be engaged in the human experience. When we want to allow ourselves to be filled with the full light and joy of the Creator, we can look beyond our normal vibrations. We can intelligently examine each belief that arises. Is there any feeling of fear? Does it resonate with the energy of the heart?

If we are completely open to what we intuitively know, and we intend to expand our conscious awareness to our actual Being, our guidance comes to us, and we can know that we are unlimited in who we are. It probably takes more than a few minutes to register that fully and expand our feelings. We can have an encounter with our own reality. In gratitude we can feel the essence of our conscious life force flowing through us constantly. By intentionally choosing to be in high vibrations as much as possible, we raise the frequency of our energy signatures and radiate our vibrations into the quantum field. They magnetically attract energy patterns that resonate with us and also raise the vibrations of humanity. We are flowing in the natural rhythm of our life current, always creating whatever we need and want. We are providing experiences for our Creator, of whom we are fractals.

In this spectrum of energy, we possess great wisdom and compassion. We feel eternal and complete on every level. In this perspective, dealing with the situations among humans becomes assisting in awakening to resolution of all of the dramatic limitations being expressed. We can change the apparent world situation with our conscious recognition of the resolutions and coming into peacefulness.

Enjoying the Freedom of Our Essential Being

Can we ever be truly free, to be unlimited in every respect? We're accustomed to living within limitations. These can be physical, such as disabilities and impairments, mental and psychological. Our limitations can be within a prescribed space or just within the perspective of humanity. The important point here is our perspective. Who do we deeply believe we are?

Our lives are circumscribed by our beliefs about ourselves. Because we are infinitely powerful creators, we can create limitations to our being. These limitations can be so convincing that we get locked into believing that we do not allow ourselves to imagine living beyond our limited abilities, including our health and our lifespan. But this is not who we are in essence.

We can dissolve the blocks in our consciousness and the limitations in our imagination and emotions. Some of us have slowly worked on opening our awareness and have resolved many emotional knots hiding within, not wanting to become known. Some of us (with a lot of Aries energy) take a direct approach and make a leap in awareness through out-of-body-consciousness and exposure to the human dimension an octave in energy vibrations beyond our Earth experience. Here we can experience a more enhanced life in every wonderful way, and any lower vibrations become unstable and dissolve, along with the beings living within them. In this spectrum of vibration the predators experience their own energy enhanced, and they terminate from within.

When we live in high-vibration energy, we realize that we are awakening from the hypnotic trance of humanity. We can still participate in the spectrum of energy of humanity from a perspective of compassionate wisdom, knowing what the experience was about. In this state of being, we can resolve any energy that faces us, bringing it into alignment with our own or neutralizing it. We are sovereign beings having experiences for our Creator and all of the quantum field and all beings arising out of it. We share our conscious awareness with everything and everyone in all dimensions. We are the personalized consciousness of the Creator. We are unlimited in our eternal Being to create our experiences.

Who Do We Really Love?

We are the personalized expressions of the consciousness of the divine Creator, existing eternally in our conscious awareness. We are unlimited in our conscious awareness and our creative ability. We live in a field of unconditional love and joy in conscious unity with our Creator. This is our true love.

We express ourselves through the vacuum fluctuation of the unified quantum field of all potentialities in our limited forms within the human experience on this Earth. We have created convincing boundaries to our awareness in order to make this experience as real as possible. Here we have been enticed into low-vibration dark energy to the point of finding out what suffering and death are like mentally and emotionally. In any other way we could not know these energies as deeply as we now do. We are gaining great wisdom through our incarnations here.

Now we have the opportunity to graduate from the lower vibrational life by changing our perspective and understanding ourselves. In the quantum field we are energy Beings, each expressing ourselves in our personal energy signature, which vibrates with the energy that we radiate through our consciousness as our aura. This is our inner design studio for our life experiences. We direct the quality of design with the vibrational frequency of our thoughts and feelings. This ability is part of our essential Being as creators and has resulted in our creation of the vibratory resonance of everything we experience.

As the energy signature of our planet rises, as it is doing now, there is a leap in consciousness-frequency to a higher dimension. Humanity must participate in this expansion of consciousness and heart energy. Our higher abilities are needed for ourselves and for humanity. We can flow with the divine energy that we are now living in and expanding with. By recognizing our potential, we can experience our natural transformation more powerfully by intentionally opening to an expanded consciousness and attracting more high-frequency experiences.

As we intuitively connect emotionally with high-vibration feelings of compassion and love, we awaken our ability to recognize the radiance of all beings and the quality of their energy signatures. Our radiance affects everyone we encounter or have feelings about. We can have gratitude and compassion for everyone we have encountered and exchanged energy with, for their participation in our experiences. We have learned the nature of a wide spectrum of emotional vibrations. Our appreciation of unconditional love is now much deeper than before the human experience. We now know this forever.

The Importance of Our Attention

How do our personal experiences come to be? They are not haphazard or accidental. Many may feel that the vibratory quality of our thoughts in our normal daily situations are not significant in our navigation of life. We are free to think about and feel however we want in every situation. What is important to know is that we are designed to be creators in every aspect of our being. What we primarily create is our life experiences. The ways that we think and feel about people and situations create patterns of energy in the quantum field that manifest as resonant energy patterns in our experiences.

Every situation that we experience is a result of energy patterns that we have registered in our attention. Because of the perspective of enslavement and poverty that humans have been taught to create for ourselves and endure throughout our lives, we have given our life force to parasitic beings who have dictated how we are permitted to live. We can free ourselves from this perspective by realizing our true Being and taking control of our emotions and our interpretation of our circumstances.

If we loo0k around at how the universe is constructed from the smallest sub-atomic waves/particles to stars and galaxies and everything in between, it is obvious that it all exists in harmony, except for the aberrations of humanity. Quantum physics has shown, through experiments that any intelligent person with access to the necessary technology can duplicate, that everything is an expression of consciousness, which is universal. It is shared by every entity that exists.

We may imagine that we are limited in our consciousness to our individual awareness, but that is because we have special talents and abilities that have enabled us to create personal limitations to our awareness to engage with all of humanity in the reality show of life in this limited dimension of energy.

We can transform our entire lives into experiences of wonder, joy and beauty just by changing our perspective and intentionally directing our thoughts and emotions into energy patterns of higher vibrations of the energy that we live within in the unconditional love that constantly flows from the universal Creator throughout the quantum field of all potentialities.

Developing a New World

We are collectively throwing off the yoke of enslavement that has plagued humanity for eons. Individually we can't change the structure of human experience on Earth, but we can change the quality of our own experiences. Prior to our incarnation we agreed to participate in an artificial reality that we all recognize as real. We have created boundaries and limitations to our awareness in order to have the full experience of fear and deprivation and all of the low-frequency energy that we live with. Now it is time for us to transform it all, drop our self-imposed limitations and open our conscious awareness to our true Being and recognize the reality of universal consciousness.

Quantum physics has been helpful in developing a rational, objective basis for understanding the nature and essence of everything we observe and think about. What has been mysterious has become mathematically provable. Universal consciousness is the basis of everything that exists. Everything is conscious right down to the tiniest sub-atomic particles, which have shown us that they know when they're being observed, can be in more than one place at the same time, and know what each one of the same wave/particles know anywhere in the universe. Their essence is beyond time and space. What does this tell us about ourselves?

Our bodies are constructed of conscious electromagnetic waves/particles. They are expressions of our conscious Being, which is so much more that we have realized. What we recognize through our attention are quantum-field wave patterns that stimulate feelings in us, and that become material in our experience. At the same time our focus is creating new experiences for us, because our thoughts and feelings resonate with energy patterns in the vacuum fluctuation. Through our focus, life force flows through us to modulate the energy patterns in our realization. The qualities of these modulated energy patterns determine the quality of our lives. As we focus on high-frequency emotions and scenarios that we create of love, compassion and inner joy, we raise the vibrations of our personal energy signatures, raising the quality of our lives to a higher spectrum of energy with more love, peace, joy, abundance and freedom.

From the perspective of compassionate wisdom, we can participate in raising the quality of energy of humanity. It is from this perspective that we, as the representatives of humanity, can create new government and finance that serve higher vibrations of experiences.

The Era of Transforming Energy Patterns

Because we are embodied in a dimension of low vibrations of fear and belief in mortality, we encounter all the challenges of humanity. These we can face from a perspective of compassion and higher guidance from our intuitive knowing. We are always free to react to everything that comes into our awareness from an enlightened perspective, knowing that we have created the quality of low-vibration energy that we encounter in order to be able to transform it.

When we encounter adverse beings and their energetic patterns, we can recognize that they have their own path in life and will respond to us according to their own free will. We can accept this with compassion and kindness, knowing that they cannot adversely affect our true Being and are holding deeply painful energy patterns from previous experience and inheritance. They feel our energetic radiance and may be stimulated to recognize that they are limiting themselves with their ego consciousness of fear and anger. The rising vibratory resonance of the regenerating Earth will realign the low-vibrational energy or dissolve it, and those who identify with it will ultimately either awaken or disappear.

When we are emotionally clear and have resolved all limiting beliefs about ourselves, we have no boundaries to our awareness and Self-realization. Our energetic signature vibrates in resonance with the energy of our Creator. We are open to our multi-dimensionality. We are present in our awareness and conscious of the creative flow of our life force as we encounter everything in our experiences. We are aligned with the flow of unconditional love and joy that is the essence of our Being as participants in the universal consciousness that constantly creates everything in our experience, formed by the quality of our thoughts, feelings and intentions.

Expanding Our Sense of Being

Although our human situation may be uninspiring because of low-vibration energy in our environment and awareness, we still have free will to focus on what we want to experience. We severely limit ourselves by focusing on experiences that do not inspire us. No matter how difficult and intrusive the low-vibration energies in our awareness may be, we have the ability to transform all that we encounter.

Our only requirement in this life is to experience being human. We don't have to be locked into any patterns of energy. We can choose to be in love and joy in ourselves as much as possible. If we can realize that we don't need any specific experiences, and are free to create whatever we prefer, we can transform everything. Energy doesn't stop flowing, but resistance to the natural flow causes stress, frustration and ultimately defeat. That's why it's important to be in the flow with the energies of nature. Flow with the fragrance of flowers, the birdsong at sunrise, the sweep of the clouds, the conscious presence of giant trees, the majesty of the night sky, the vibrancy of a coral reef, the beauty all around.

If we can just be present in the energic spectrum of the Earth, we come closer to our true Being. We can realize that we control the quality of our lives just by how we feel and what we think about. Our state of being determines the vibratory pattern of our personal energy signature. This is the quality of energy that we radiate into the quantum field, where the conscious, creative plasma energy forms experiences for us in resonance with our energy signature. We magnetically attract persons and experiences that resonate with our energy spectrum.

By intentionally elevating our feelings and thoughts, we modulate the energy in our presence into resonance with higher vibrational experiences. If we feel that it's too difficult to hold that perspective, because of the intensity of our poverty or enslavement, we can naturally repel low-vibration scenarios by intentionally aligning with the energies of nature, inspiring music and art and any environment that we can find that may elevate our feelings. Inspiration comes from our own Being, and we can be open to it through promptings from inspired creations in nature and by artists and poets. And we must ask for it and intend to be aware of it. Our awareness can transform our limitations through our perspective of high-frequency emotions and thoughts of joy and compassion These naturally come through our heart and are in the flow of our natural life force and that of the Earth.

Living in Harmony or Conflict

Love is high-vibration, life-enhancing, connective energy. It connects us to each other and to all that exists through universal consciousness. Fear is low-vibration, life-destroying, disconnective energy. It separates us in our conscious awareness from each other and from all that exists. In our own being, how we perceive, react and create with these qualities of energy determines the vibratory frequency of our personal energy signature, resulting in the quality of our experiences.

We live within a dimension of energy that we have created to confine our awareness to the vibratory spectrum of the human experience. There is a veil of energy that separates the energies of love from the energies of fear. If we live in the vibrations below this veil, we feel fearful in some way and always in fear of mortality. We have designed this dimension so that we cannot experience the unconditional love that fills universal consciousness, uniting all conscious entities throughout the cosmos and beyond.

By clearing out the limitations that we have imposed upon our awareness, we can open ourselves to living in the realm of higher consciousness, where life is wonderful. We designed our false beliefs about ourselves to be imbedded in the depths of our consciousness, so that we could not be aware of them. We can know what they are whenever they appear. We can feel the emotional knots of low vibration that bind us in certain situations. There's always some level of fear, which indicates to us that we're in the realm of low vibrations and need to transform the situation in our awareness by knowing intuitively that we are eternal Beings of pure conscious awareness. In our presence of awareness we have unlimited ability to create universes and whatever else we want to create in love and joy. In the realm of fear, we could not even imagine how this could be true.

We can pay close attention to our emotional awareness, because it feels the quality of energy that we focus upon. This is our guide through our human experiences. We are not limited to low-frequency lives. We have allowed them to be created for us unknowingly, but we have the free will to choose a higher vibratory life at any time. This is how we can use our emotions creatively.

Making the jump from fear to love, from feeling separate and mortal to feeling eternally connected to everyone, requires a drastic change in perspective. We can know this intuitively, but our deepest beliefs may not allow us to realize the truth of our Being. We can transform those beliefs, first by awareness of our intuitive knowing and feeling, then by mental acceptance and understanding, and eventually by experience, especially if we yearn for it.

Transforming Life on Earth

We are the masters of life here. We are the ones creating all of our experiences. No one outside of our own consciousness has any power over us. Our disempowering experiences are all a result of the expectations and beliefs that we use to give our power away. By taking on fear for survival and feelings of victimhood and poverty, as well as anger and dissent, we disempower ourselves by creating what we do not want. But this is why we're experiencing these conditions, so that we can learn how to work with the energies in our being. We are entrepreneurs in the adventure of living in a contrived Matrix of energies. We are experimenting and challenging ourselves to find out how everything works.

Eventually we learn that nothing outside of our own presence of being has any power over us. We express ourselves through our energy signature. We can live in our aura, in a high vibratory state of being, inaccessible to low-vibration energies.

Our magnetic polarity repels them. They are not present for us. They have no reality, except what we give them. They are energy patterns that flow past us without interference.

It is our choice in every moment to focus upon and feel any imaginable energy patterns that we choose and the consciousness that supports them. This focus expresses the quality of creative energy that we align ourselves with. It is this vibratory pattern that attracts other vibratory patterns that resonate with it and repels those that don't. These activities and interactions of energy patterns have been intuitive for us, but they have been recognized as mysterious. They are actually clear in quantum physics.

The only transforming we need to do is within ourselves. Our experiences will resonate with the vibratory patterns we focus upon. Our entire perceived environment resonates with our own energy signature. We can accept our entire environmental experiences from a creative perspective. They are projections of our own conscious awareness. We can play with the vibrations of our intentional awareness, creating new and wonderful scenarios. As we experience these scenarios, we learn to recognize how everything feels, and we can direct our emotions to the vibrations we choose.

To the extent that we include humanity in our awareness, we can stay in high vibrations emotionally and intentionally, raising the vibrations of the energy signature of humanity. We can witness everything from the perspective of compassionate wisdom. All of our interactions can begin to align with love and joy, and we can live in abundance and freedom for all in our intentional awareness. This can be our present and our future.

Becoming Aware of Our Expanded Being

If we want to awaken from the hypnotic trance that we have lived within for eons, we must strengthen our resolve to do so.

Chapter 1. The Power of Our Perspective

We continuously create the energy spectrum that we live within. Only by expanding our creation can we open our awareness to a greater sense of being. We need to raise the vibrations of our energy signatures. If we've explored the dark, low-vibration realm to our satisfaction, we are ready to leave that behind us. We do not wish to expand our awareness into the depths of low-vibrational experiences any more intensely. We've learned great compassion and wisdom. Now it's time for us to live in love and joy in our eternal presence. If we want to expand our consciousness, we must go higher in vibrations of all kinds without fear.

High-vibration energy feels good. The higher we go, the better we feel, and the more optimistic we become. By focusing on visions and emotions in a higher spectrum of frequency, we raise the vibratory rate of our energy signatures. As a result, low-frequency energy is not attracted to us, and we are magnetically polarized oppositely, once we rise above fear and enter the vibratory level of goodness.

All electromagnetic wave patterns that we can be aware of stimulate emotions in us. This is how we know the frequency level of the people and situations that we are facing. Our emotions have not been subject to the mental conditioning that we have endured. They still feel the wave patterns and frequencies the way we were created to feel. In even very subtle feelings our emotions are aware of the distinctions between fear and love. This can be our path to a higher dimension of living, closer to our natural expanded Self.

We are absolutely free to choose what we focus our attention on and how we want to feel in every moment. In the midst of great trauma and chaos, we can choose the be in compassion and understanding. We can focus on the inner light in everyone and everything. It is always present, or whatever embodies it would not exist. We can consciously invite it to show itself to us. This is the only energy we need to relate to. This perspective opens us to greater awareness and expanded consciousness. Ultimately it carries us into the universal consciousness of our Creator.

Our Eternal Self-Aware Presence

If we are each an eternal presence of consciousness, what really is that? We have learned that we are our ego consciousness, localized in our bodies, but we can expand our awareness far beyond the consciousness of our body and ego. We set our own limits, beyond which we are uncomfortable about imagining experiences. We are fearful and disdainful in explorations in consciousness. The shift from fear to joy is challenging, because the ego can't do it. It requires dropping all of our beliefs about our personal and collective limitations. It can happen when we open to our intuitive guidance and knowing. We learn to trust ourselves by becoming aware of our inner knowing. It is who we are apart from every conception we have about ourselves.

It is beyond the ego. Our egos are aspects of our consciousness that allow us to function within a spectrum of frequencies characterized by density of energy in physical matter. It is also the realm of fear and all of the low-vibrational emotions and thoughts, such as poverty, greed and desire for power over others. The ego has become our navigator and guide without our intuitive knowing. It is blind to our higher guidance, but it has been necessary for us to have the full human experience. We have become so accustomed to our limited awareness, that we do not even recognize that we have higher guidance. Most of us cannot even imagine that we can change the circumstances of our lives just by changing our perspective and beliefs about ourselves. The transformation doesn't require time, because it happens in the quantum field outside of time. It's completely a matter of modulating the frequencies of energy patterns in our awareness.

As we learn to be in alignment with the vibrations of our intuitive knowing and compassionate wisdom, we may face challenges from low-frequency situations that appear for us. We intuitively know what to do, because the vibrations are part of our own consciousness. We can modulate the energy vibrations with our recognition, while feeling the presence of unconditional love all around and within us. This lifts our consciousness into a higher dimension beyond fear and into love. Our experiences come into alignment with our personal energy vibratory spectrum.

By being aware of the energy of our heart through our emotions, we can learn to focus on living in high-vibrational scenarios as Beings of light and love. This is our inherent nature, and it's what we're being attracted toward. In our personal eternal Self-aware presence, we are unlimited conscious awareness, abiding in gratitude and joy, aware of our unlimited creative abilities with our attention and emotions.

The Natural Flow of Life

The tractor beam to a higher dimension of living is drawing us into feeling higher vibrations, living more loving lives. This is the direction of the flow of life force that enlivens us. We can observe this in the Shumann Resonance graph of the Earth, where we now see four higher octaves of vibrational patterns raising the vibratory frequency of the energy signature expressed by the consciousness of our planet. The natural flow of life force is encouraging us to elevate our life expressions. We are being drawn into alignment with the rising vibratory rate of the Earth.

We are being called to be in nature, walking barefoot on the Earth to ground our energy and receive the rising energy radiance of Gaia and the natural radiance of our Sun. Here we can be peaceful within, breathing the vibrant air deeply and being open to feeling the energy patterns of the Earth. We do have this ability, although our perspective, containing all of our beliefs, filters our understanding of our situation. Indigenous people have always aligned themselves with the energy of the Earth. This is natural for us, once we resolve our conditioned beliefs of limitation. We know this energy intuitively. We just need to be open to realizing it.

These rising frequencies of the energy signature of the Earth and humanity are influencing our lives in every way. Everything and everyone living in the low vibratory spectrum of fear and limitation is being challenged to awaken to a higher spectrum of energy. If they don't align with the flow of higher vibrations, their energy signature becomes unstable, adversely affecting conscious functioning. Everyone and everything living in the high-frequency spectrum of love and compassion is being supported to go higher as well. This is following the natural flow of life at this time.

We are moving into realizing that we are all sovereign Beings, eternal in our essence and expressing ourselves in the recognition of our human bodies. We have control of the quality of our lives through our mental and emotional expressions. The vibratory patterns of energy that we are holding in our awareness in any moment attract vibratory patterns in resonance with them. This results in our experience of similar qualities of energy in our future.

The Role of Starvation, Disease and Suffering

In this lifetime, we have come to complete our low-vibration Earth experience and ascend into a higher dimension of frequencies. Our destiny is full awareness of our eternal Being and the infinite love constantly flowing to us in the life force of our Creator Consciousness. In our essential Being, we have always lived in the higher vibrations. This is our natural environment.

Our journey into materiality and a low-vibrational experience is ending. The rising frequency patterns of the Earth and of our entire cosmic environment requires our alignment with higher vibrational living. Many of us have fallen into difficult circumstances through lack of alignment with higher vibrations. We've been intimidated into accepting low-vibrational experiences. We have accepted lives filled with stress, and always in the background is fear. This self-deprecating vibration attracts resonant experiences of hunger for food, better health, love and better circumstances. If we focus on experiencing low-vibration situations of lack, because that's what we're already experiencing, we continue to modulate the energy patterns around us into the circumstances that we are emotionally involved in.

We're being challenged to change our perspective, regardless of our circumstances; otherwise we'll continue to create worse situations for ourselves to experience. When we're really starving, deathly sick or being hunted down, our ego becomes very disturbed, to the point of becoming unstable and desperate. It knows it's heading toward termination. The only viable alternative is transformation. Give up. Breathe. Be aware of how we feel. We can breathe deeply and rhythmically until we feel neutral and receptive to our inner prompting.

Whatever state of being we occupy is subject to our intentional presence of awareness. We have absolute freedom to choose what vibratory patterns we want to experience in any moment, regardless of what we are confronting. We can use our imagination to create scenarios of beauty, gratitude and abundance, and we can feel ourselves living in them. We can create virtual reality with our imagination and emotions. If we don't interfere with our visionary experiences by disbelieving their potential creation, their quality of vibrations will become the quality of our experiences. Our natural energy is filled with vitality and joy.

Finding Our Way Energetically

Since infancy we've accommodated the perspective held by most humans. This perspective includes belief that we are imperfect in many ways, vulnerable to intimidation and mortal. Our personal experience bears this out. We believe that reality is limited to empirical experience and physical expression in our sensory perceptions and in the world outside of our body. But what if all of this is just a projection of our own consciousness in conjunction with everyone else? How can we know what is real?

In quantum physics everything is an expression of universal consciousness. Every expression has an electromagnetic energy signature that resonates at a measurable frequency. It's unique to everyone. The resonant frequency of the Earth has been measured at 7.83 cycles per second. In order to live on this planet, we must resonate in alignment with the vibratory frequency of the Earth. Our energy signature contains the patterns of all of our thoughts, emotions and beliefs, from the most deeply held to our current passing fantasy. Everything that we experience in the empirical world is a complex of interactions of energy patterns that we recognize and interpret as physical experiences. There's nothing solid about the empirical world. The amount of condensed energy in all of the atoms is miniscule, compared with the space within the atoms. What we perceive is the spin of the energy patterns that we interpret and recognize as solid material.

Physicists have shown that everything is energetic patterns of electricity and magnetism. Our bodies are electromagnetic expressions of our personal consciousness. Everything we experience is a projection of our consciousness. Universal consciousness is constantly creative, and we are part of it. We participate in the creative process by recognizing energy patterns and changing them with our thoughts and emotions. We are modulators of energy. Our thoughts give form to our experiences, and our emotions provide the quality or feelings of our experiences.

By recognizing ourselves as pure eternal consciousness, we can create any life experience that we want. We are unlimited Beings with infinite abilities. We are constantly creating our experiences in every moment by the quality of our presence. We all know this within our own consciousness. To gain access to it, we must resolve all of our fears and limiting beliefs about ourselves and learn to live in love and peace, even if we have to pretend for a while.

A Possible Perspective on Life

I AM that I AM. What does this really mean? It is realizing our conscious life force flowing through us. It is the Creator Consciousness eternally flowing through us. It is our Universal Consciousness, our awareness of our divine Being. It is who we are in our true essence. We are the same conscious Being as the Creator. Each of us is the infinitely creative Creator. In our complete Being, each of us is unlimited in every way. We are the Supreme Creator's consciousness. We provide experiences of all kinds for the Creator Consciousness. The Creator receives our thoughts, emotions and situations in every moment.

We vibrate at the frequency spectrum of the focus and feel of our attention. We choose our focus and how we feel about it. We also feel what's going on in the quantum field around us within the focus of our attention.

We can focus on any vibration that we choose. If we let our emotions freely guide us to higher vibrations, we can imagine things and interactions that we feel really good about. We can look at every situation that we confront from a perspective of gratitude, compassion and love. We cannot be threatened without our permission, because we are the Infinite Creator in our expanded consciousness. Knowing that we are infinitely creative, we can be in any challenging situation and transform the energy to higher vibrations. Even if we don't know that we can do this, we can practice, until it becomes real for us.

If we are poverty-stricken and starving, we will be very challenged to imagine with gratitude, love and joy, our next wonderful meal. But we can do it as much as possible. If reasons appear in our awareness of why we shouldn't or can't have that meal, they must be resolved through compassionate wisdom. Eventually we'll be successful, and the miracle can happen and keep happening, if necessary. Meanwhile we're not starving, we're fasting, because that's actually what it is. We're preparing to be receptive to our heart's intuitive guidance. We're releasing all of our extraneous energy from low-vibration parasites and becoming clear about one thing—our survival. We are being asked by destiny to recognize our eternal Being. It's time to awaken from the hypnotic dream of the human Matrix into realizing our true presence of Self.

Rising Earth Resonance Alignment

As the resonant energy frequency of the Spirit of the Earth continues rising, the lower vibrations are becoming unstable, having cut themselves off from most of their life force through their consciousness blocks held in place by their powerful belief in evil. The dark ones are losing their life force and are desperate to survive. We can be compassionate with them, while we maintain the perspective of our highest vibrations and radiate love and wisdom. We invite the energy of every person or circumstance to come into alignment with higher vibrations or disappear from our awareness, as we withdraw the support of our life force from low-vibration energy patterns. Any frequency not in alignment with love and joy is being transformed. The vacuum fluctuation of the quantum field is moving in higher frequencies. Lower-frequency energies cannot withstand the change and must either come into alignment with elevated being or disappear. Their only reality is the focus of our attention. Without our recognition, they cannot exist, except as passing wave patterns in a dimension different from ours.

Humanity is fixated on the empirical spectrum and is in a hypnotic trance, but doesn't realize it yet. Working with quantum energetics is semi-empirical. It observes how the smallest empirical particles act, and has shown that they display infinite awareness, and they operate outside of time and space. This means that we're made of materials that don't exist in the world that we perceive. It negates the solidity of our material world. What's left of the empirical world is our belief in its reality.

We can transform our personal experience, regardless of what situation we're in or enveloped within. We can be guided by our intuition, which is what we most deeply know without any fear, only love. This is where our love is true, with gratitude and joy. By holding this elevated spectrum of vibrations, we are modulating the energy that we confront in our awareness. This results in elevated experiences for us, regardless of what anyone else may be experiencing.

We are being freed from third-dimensional energy. We only need to open our awareness to higher frequency feelings and visions. We know when we're expanding, and this is the energy spectrum that we're being attracted to. We can be free of personal drama and insecurities. We can stop recognizing them. We innately know love and compassion deep within. Without fear of any kind, we are naturally expansive as our vibrations rise. We can become brighter and more radiant. We can be aware of our constant creative ability, and we can use the power of our focus of attention to modulate increasingly higher energies, which we know through our emotions.

The higher we go, the better we feel. It's all relative to our perspective. We can be resolute in our increasing vibratory spectrum, always being aware of the energy of love and recognizing it in everyone. This is the energy that fulfills everyone and everything. We are expanding far beyond humanity's current energy spectrum. It works by constantly feeling better in gratitude for our self-aware being.

Releasing Our Fixations

To those of us who are stuck in poverty or enslavement of any kind, we have the conscious choice to redirect our energy from our fixation on these things. If we can continue to deal with life as it confronts us and instills expectations, we can choose to be in a perspective of compassion and heart-felt knowing. This perspective transforms a low-frequency fixation through understanding and realignment with the high-frequency energy of inner joy, freedom and abundance. By holding visions and feelings of high-frequency experiences, we direct our life force in creating what we are passionate about in our deepest Being. This transforms our condition in the world from one based in fear to one based in love. It occurs on a vibratory level that we direct with our focus of attention and emotional determination.

We live in a cosmos composed of infinite patterns of electromagnetic waves that can appear for us when we recognize them. The material world appears in our experience, because we recognize its patterns of energy. By changing our perspective and our focus of attention, we can create a different personal quality of life. We can recognize the energy patterns that we want to experience, instead of the ones we don't want. We can reclaim and redirect our life force, as we move deeper into the understanding and feeling of our heart. We become much more aware of everything and more capable of dealing with the situations of humanity. We are on our way to awakening to who we really are.

We know from the accounts of people who have died and come back, that our self-awareness of personal identity does not change at death. We are eternal, self-realized persons. This is the basis of our true perspective on life. Once we have this realization and know deep within that it is true, we can release our fear of suffering and survival. These have been boundaries in our consciousness, keeping us from complete Self-realization. We can resolve our fears, opening our awareness greatly to high-vibrations in expansive consciousness. This is completely life-changing.

The Essence of Truth

We live in a world of pretense. There are so many lies, so much propaganda and so much suppression of the truth, that we face a great challenge of understanding what is real. We know from quantum physics that everything is energy and vibration. Everything that is, except consciousness, which expresses itself as energy and modulates the energy into patterns of vibrations in the quantum field. Because we are conscious beings, we can align with these energetic patterns, and in the spectrum of energies recognized by humans, the energies manifest as material substances. Each substance, when not observed, vibrates in its own spectrum or energy signature as electromagnetic wave patterns. We can feel these wave patterns and focus on them, causing them to spin instantly into materialization. When we no longer focus on them, they turn back into their signature energetic wave patterns.

The energies that we focus on in the current human experience can be complex. Much communication consists of falsehoods that are presented as truth, but their vibrations are low, in the spectrum of fear. We can feel the vibrations. We know when fear is present, and whenever it is, and whatever is being presented cannot be true. Only communications sent in love can be true. The communicator must want the best for all who receive the messages. We know when this is true, because we can feel it. The vibrations stimulate goodness and compassion in us.

If we do not get dramatically involved in material forms and expressions, we can pay attention to the vibrations that are present for us. They can tell us what is true and what is not. We feel them, and we know. It's all part of our own consciousness. This goes for our living conditions as well. We know what is true for us, and what we have accepted instead.

We do not have to stay in any spectrum of vibrations. We have the freedom to focus our thoughts and feelings on any experience we are interested in. If we want to change our life from being enveloped in low-vibration energy to something better, we can change our perspective to being in a high-vibration environment. We can awaken to each day in gratitude for our presence of Being and for everything we experience, because it is all for our personal enhancement. We can bring all of our own discordant energies into alignment with the energies of our heart. There need be no negative feelings of any kind from the past, present or future. We can leave all fear-based experiences and intentions behind and no longer focus on them. By maintaining our focus on high-vibration experiences, we can transform our lives and move into a higher dimension of reality, where only truth resides.

Our Present Self and Our Potential

Our ego cannot know truth. It can only believe what appears to be truth in the low vibrational world of humanity. There is no truth in the lower vibrations, only pretense and trickery. This is becoming more obvious day by day. Since fewer and fewer people believe what presents itself as truth, the pretenders are becoming bolder and more desperate to survive, as they become unstable under the influence of the rising resonant frequency of the Earth, as shown in the Shumann Resonance Graph. The vibrations of Gaia are now vibrating strongly on three higher octaves of vibrations beyond the traditional 7.83 cycles per second, and a fourth is faintly beginning. People oriented to low vibrations, based in fear and greed, are having more volatility in attempting to maintain their frequency patterns. Gaia is becoming less hospitable to low vibrations. As she rises several octaves in frequency, all life forms on our planet will be brought into alignment with her, or they will become unstable in their frequency, and their amplitude starts being suppressed. Their life force begins to fade, as they close themselves off from a perspective based in love, which is the frequency Gaia is at home in. In this frequency, there is only truth.

We know when our sense of fear is being stimulated by low vibrations. We're not sure about true love. We may not ever have experienced it, because it vibrates in such high frequencies, that it is beyond our human limitations to imagine and feel. We must go beyond the ego and our limited human-conscious mind. We can begin this journey through advanced breathing techniques, deep meditation, advanced yoga, Sufi trance dancing and many other advanced spiritual practices. This is serious work.

We can analyze the nature of what we have comprehended of the cosmos through quantum energetics, and we come face to face with the mystery of universal consciousness and the unified quantum field of all potentialities, which envelops us. We all arise out of universal consciousness, which expresses itself in the Being of each of us as a fractal of Itself. When we align with the frequency of this energy, we realize who we really are and what we're capable of.

We can clear ourselves of all of our beliefs about ourselves, and we can love our ego into allowing us to make a leap of faith into the higher-spectrum frequencies of life beyond time and space. This is the realm of magic and wonder. It is the vibration that Jesus referred to, when he showed us what is possible for a Being with a perspective of expanded consciousness and told us that we can do the same things he did, and even greater. It requires our alignment with the highest-frequency energies we can imagine and a willingness to live in this energy spectrum and express ourselves through the guidance of the energy of our heart. In this way we can create our lives in the new world of higher vibrations.

Vibrating into Universal Consciousness

Quantum sciences have identified universal consciousness as the source of all that exists. Science has not been able to quantify this consciousness entirely, because consciousness exists in every dimension and plane of existence. It is not bound by the limitations of science or materiality. And yet we can know its essence, because it is our own essence.

Traditional human understanding of consciousness has a resonant frequency limited by fear. We fear pain, suffering and the threat of termination of being. We create these threats that stimulate fear, in order to realize that they have no essence apart from our own fear. Our fears give our life force to these threats, which attracts energy patterns that align with our fears for us to experience.

By intentionally changing our perspective to higher frequencies, we stop creating fearful experiences. We become able to open our awareness in every moment to unbounded, joyful realms beyond our imagination. We can use our imagination and creative emotions to stretch our reality toward realizing our timeless essence of Being. We reach for the vibrational spectrum of everyone and everything that we love the most.

We are Beings of light. We glow with a radiant aura of photons empowered through the light of our heart. This is our natural essence, along with our emotional alignment with unconditional love. We become able to be aware of the high-frequency life force of the Creator flowing through everyone. We may choose to interact only with the light of the Creator in all our encounters. We know the messages of this inner light in our intuition. We can perceive the world of humanity from a spectrum of vibrations in a higher dimension, where there is only love and compassion for the human drama.

We can jump beyond our traditional awareness into our eternal Self-consciousness in any moment by being completely present in our awareness without limits, as we open ourselves to our natural state of Being, vibrating through the energy of our heart.

Recognizing Each Other in Ourselves

We naturally encounter each other through our eyes. We receive each other's vibration and feel each other's presence. Living within the energy that we radiate, we attract resonant frequencies in each other. We can feel the vibratory resonance of each other, and we can recognize our perspective on the love/fear vibratory patterns. We can find the Source life-force flow in each other. It comes through the vibrations of our heart. It is our inner light, our intuitive knowing and our Self-realization. All of this we communicate through our eyes and our presence.

We know immediately if we're in alignment vibrationally with anyone we encounter. Because we attract energy signatures that resonate with our own, we are facing ourselves reflected in each other. It is the quality of the energy present, the vibratory spectrum, that we encounter. The persons present and the situations may change, but the vibratory spectrum of energy determines the quality of every experience. There is a dimensional leap from the low-frequency spectrum of fear and the high-frequency spectrum of love.

The ego lives in the dimension of fear and without higher guidance. We created it for this purpose, to be able to deepen our compassionate wisdom, having experienced fear. We can now ask our ego to become an objective observer, while we can intend always to be aware of our intuitive knowing, which we can have in the dimension of love.

Our experiences are reflections of our own vibratory patterns. We always face our own energy patterns in every moment. If we want to elevate the quality of our encounters, we can align our perspective with the life-giving energy spectrum of our heart. We can be thankful and joyous to just be present in timeless awareness without limits. We can recognize ourselves as our personal presence of Being, distinct from all other persons, and all of us constantly arising out of the One universal consciousness, which is the essence of our eternal Being and the source of our knowing in every moment.

Recognizing Ourselves as Inspired, Luminous Beings

Every cell, every atom and every sub-atomic particle/wave pattern in our bodies is consciously alive. All naturally align vibrationally with the life force that streams through the heart of our Being. Since they are all also part of our conscious Self, they align also with our personal energy signatures. When there is misalignment within our consciousness, due to the presence of fear, our physical cells become vibrationally misaligned, and our bodies' life processes begin degrading.

If we can recognize ourselves as the directors of our own universes of trillions of cells, we can be responsible for inner vibratory alignment with our natural Being. We are constantly enveloped within the quantum field of universal consciousness and are given the choice of controlling our attention and focus, in order to align with the energies coming through our heart. The universe of cells within each of us has a natural tendency to convey the flow of life force in alignment for the greatest vitality of all. We can consciously enhance this flow of vitality by recognizing what it is and learning to direct our focus to it. It is formless energy in universal consciousness sustaining everything and everyone. This energy flows through our Being, connecting us in our essence to all that exists.

Our attention can include awareness of the qualities of the creative life force that we embody. Its high frequency vibrations stimulate feelings of gratitude, love, joy and all compatible emotions. We can become aware of great beauty and music and the high-vibrational abilities of telepathy, clairvoyance and more. By aligning with the greatest vitality, we are resonating with the natural flow of our life force.

This is the flow of energy that we can modulate with our thoughts and emotions in our attention. In every moment we have the choice of focusing our attention. We can choose to focus on high-vibrational scenarios and persons. As our energy signature rises in frequency, we feel more empowered and enjoy greater vitality. Our cells operate with greater vitality. We release more photons and become more radiant. We can begin to know unconditional love, connecting us through our life force to all that exists.

Aligning Our Resonant Vibrations

We are designed to be absolute masters of our experiences. If we can know what it feels like to align consciously with our energy signature, we can recognize the quality of our vibrations immediately. In every case, where there is fear, we can face this with compassion and love. By changing our perspective from fear to love, we can enter a higher dimension of resonant frequencies. By expecting and knowing a range of emotions, we attract experiences, energy patterns, that are compatible. We create what we expect to recognize.

We can choose the vibratory patterns and frequencies that we focus on in every moment. By staying in the love spectrum, we can resolve everything into universal consciousness. We can become our present awareness without limitations modulating everything with our imagination and feelings into alignment with the natural vibrations of our heart energy. In this alignment we feel fulfilled, knowing that we are creating whatever quality of experience we focus on.

Our experiences all result from the vibratory frequency of our energy signature, which is controlled by our emotions. As we go through life, we are presented with many situations. Our reactions in any situation contribute to the vibratory frequency of our energy signature. If we can stay in the realm of love and compassion, we can maintain a high-frequency energy signature, attracting high-frequency experiences and participating in every encounter as guided by our intuition.

In this spectrum of frequencies things happen that seem extraordinary, because we're working with energetic frequencies that can change moment to moment. By changing our perspective from fear to love, our entire spectrum of energy changes, creating a new life in a higher dimension of vibrations.

Mastery of Energetic Dimensions

We are participating in a game of consciousness. In our current dimension of human experience, we have compartmentalized our awareness to be able to experience fear in all of its aspects, which we could never know in our true Being, in which we know only unconditional love and joy. We had to impose false beliefs about ourselves, and we had to erase all memory and awareness of our natural essence and abilities. This game is now ending, as the Earth enters an expanded phase of consciousness. Humanity is awakening in recognition of the limitations of consciousness that we have experienced as reality. Our limitations have been based upon a spectrum of low-vibration energy that expresses fear, but the vibrations of Gaia, the Spirit of the Earth, are rising, along with the entire energetic spectrum of our enveloping cosmos. This is destabilizing the low-frequency vibrations and everyone who aligns with them and is creating greater vitality and brilliance all around and within everyone who resonates with the rising frequencies.

It is not part of human training and programing to recognize that we are energy Beings. As a result, we have been unaware of our capabilities and have believed that our lives are directed by fate or chance or punishment and reward. We have been taught that everything that happens to us is caused by forces or beings outside of ourselves. Someone else has caused us to experience pain and suffering, or we've had a stroke of good luck. This is not how life works. Everything we are and experience arises from our own Being and is reflected outwardly. Our perspectives and beliefs become our experiences.

Everyone who has had an out-of-body experience has reported that when we, in our conscious awareness, leave our body, we enter a realm beyond time and space in a different dimension. It's not necessarily a different place, but it's a different vibration. It is less solid and is instantly created by our imagination and emotions. It is filled with wonderful persons and imbued with the unconditional love and joy of our unlimited natural consciousness. All who have died and came back have reported that they were so fulfilled that they did not want to come back, but did so, because they knew that they had not yet fulfilled their embodied destiny. They were forever changed and no longer were subject to fear, because they had experienced our eternal Being.

We do not need to die to experience our eternal Being. We have been created to be Self-realized in the fullness of greatest love and creative ability. Even in our limited state of being human, we still are our true Being and have access to our ability to modulate the energy patterns that we encounter. We are the Creators of our lives, and we are now awakening to our truth and are beginning to cooperate in creating the beautiful new world that we all want to participate in.

Understanding Our Beliefs

Living in the world of human experience without higher guidance prompts our acquisition of beliefs that we learn in our interface with others. We allow them to tell us who we are, how we appear, what we are allowed to do and think. These are all limitations on our realizations about life. This is a world of ego consciousness, which we constantly create in our attempts to avoid threats to our diminishment or termination. The ego is an artificial self that lives without a conscious connection to the Source of all Being. The ego believes it is a completely separate being in an empirical world where encounters and events transpire. The ego must figure things out, make plans and take precautions to ensure existence. Egos seek comfort with families and friends, whom we trust, and which we may feel enable us to extend our presence in memory and influence beyond our limited lifetime.

In each moment we choose the focus of our attention and our perspective. If we focus on low-vibratory experiences from a perspective of fear, we are consciously aligned with that energy and we attract experiences that stimulate fear in us. If we choose to focus on experiences that are fulfilling for all, we attract experiences of fulfillment in our lives. In the quantum field we magnetically attract energetic patterns that resonate in alignment with our own. These become our experiences, and we will choose the perspective from which we encounter each of them. Our vibrations in these encounters create the energetic patterns that are reflected back to us as experiences in the world. Within our own consciousness we are constantly creating the qualities of our lives by projecting our life force through our imaginings, feelings, words and actions.

All limiting beliefs fall away from our awareness, once we open ourselves to the energy of the heart of our Being, where we can feel and know the quality of our life force. Symbolized by our heart, this energy constantly streams into us from our Creator's consciousness. It is our consciousness, and it is everywhere always creating everything. It is also nowhere or now here. In our essence we are our clear, present unlimited awareness, and we can align with the rising vibrations of Gaia, our Earth Spirit. These are all life-enhancing energies of love vibrations and are the vibratory frequencies that align with our expanding awareness of a higher quality of life. It is into the spectrum of gratitude, love, joy, compassion and forgiveness, that we are being drawn, if we are open to it.

Living as Transformers of Life

One way to transform our lives is to understand the energy that creates all experiences. We live in a field of interacting energies of all frequencies and patterns. We express ourselves through the ones that we recognize. These are the ones we choose to pay attention to. Our attention sends the life force that we modulate as it flows through the heart of our Being to materialize the quality of our energy patterns as our experience.

We have been trained to react to situations that we encounter. In our reaction, we magnetize energy patterns that align with our vibrations during the encounter. It is the vibrations of our reaction that create our experience. We could react from ego consciousness or from intuitive guidance. If we react to a perceived attack from a perspective of fear or aggression, we attract experiences in alignment with our perspective. Because of our energy modulating ability, we are constantly creating qualities of energy patterns that manifest for us out of the quantum field. We can use our challenging experiences to empower our creative ability to a higher level of vibrations.

Nothing happens to us without our creative permission. When we are threatened, it is because we have focused our attention on that quality of energy. If we become intuitively-guided, we will know that there are no threats to our real Being, and we can encounter everyone and everything with gratitude, compassion, love and joy. These energetic expressions attract energy patterns that stimulate feelings of love and joy as they manifest for us.

The quality of life that we believe in is the quality of life that we experience. Our beliefs create the expressions of our consciousness. Changing our beliefs changes the corresponding quality of our lives. Ultimately we'll transcend all beliefs as we enter an awareness of universal consciousness, and our creations arise from a perspective of clear unconditional love and joy, which flows to us in each moment from the consciousness of the Creator and enlivens our eternal Being.

Our Evolving Transparency

In times past, personal privacy was highly regarded. Today we can still have our own thoughts in private, but low-vibration thoughts and emotions are even broadcast to us by advanced technology programmed by the beings who control humanity. This is done not only by subliminal messaging in all major media, but also via powerful energy beams. It has become challenging to focus on uplifting energies, especially in cities with concentrated populations. We do, however, always have freedom of choice about what we are aware of and what we want to dwell upon and create in our experience.

Each of us has an energy signature that is constantly vibrating within the spectrum of our accustomed vibrations of thoughts and emotions. Our technocrats are building artificial energy signatures for each of us, based on our conversations and comments online, on our phones, all the websites we've gone to, all the things we buy online, all our sources of income and everything we own. They already know all of this and more, and it resides in banks of supercomputers. All of this data has a dig-

Chapter 1. The Power of Our Perspective

ital energy signature for each of us. By recognizing this energy signature, the AI system can devise treats that align with this energy signature. These can be used for temptation to go into lower-vibration situations and stay there. These are all based on some kind of fear. Fear sends our life force to enliven the creators of the fear. Without our recognition of their threat and its stimulation of our fear, they could not exist. They can also cause feelings of inhibition for higher-vibration experiences by instilling low-vibration beliefs into our consciousness. These are things like shame, hatred, aggression, superiority, inferiority and limitation.

As we begin to become aware of the magnitude of our consciousness, beyond time and space, privacy in any form is impossible, because we know everything, whatever we focus on or inquire about. We are transparent in every way. For anyone who is in fear, this kind of transparency is painful. It's not synchronized with our intuitive knowing in the energy of the heart of our Being. Low vibration energy has its frequencies interrupted by the high-frequency energy constantly radiating from our heart, when aligned with the essence of our Being, as we arise out of the universal consciousness of the Creator. Low-frequency situations cannot exist in our experience, when our energy signature continuously vibrates at high frequencies. We do not need to recognize low-frequency situations as such. From an expanded perspective, they transform into experiences of compassion and love. This is how we can choose to recognize every encounter we experience. Our personal lives become filled with magic, beauty and abundance.

Being Open and Aware Beyond Beliefs

As long as we believe that we are no more than the body and personality that we live with, we cannot know anything more, unless we become so uncomfortable with our accustomed lives, that we begin to see hints of a better life. This situation is now establishing itself among humanity, because we are on the verge of complete enslavement and impoverishment world-wide to a ruthless group of low-vibration controllers, who are themselves feeling pain and discomfort in having to live with the rising resonant frequencies of this planet.

Finding out everything is energy, and nothing is substantial, leads to potential awareness of different frequencies and patterns of energy than we usually allow ourselves to encounter. We have the free choice of deciding where to put our attention and what we want to feel. The choices available to us are as great as our imaginations can be open to. We can get stuck with choices of forms, such as money or food or a new cell phone, and miss the important creative element, which is vibratory frequency and resonance. These characteristics of energy determine the quality of our experiences and whether they are beneficial.

We have the ability to transform energy patterns with our intentional will power. When we are clear and centered in our perspective, we wield potentially powerful creative radiance into the quantum field for manifestation in our experience. When we encounter any situation, we have the choice of how to receive it or react to it. If we align with the energy of the situation, we create more of it in our experience. If we can identify a situation from a perspective that is open to feeling the energy patterns that we face, we can choose to receive them as gifts, and that's what they will become for us.

There is some measure of amplitude in the vibrations of every energy pattern, even very low-frequency ones. We can feel the quality of all vibrations that we allow into our awareness. Any energy patterns expressing low vibrations feel fearful at some level. When we realize that we are encountering fear, we can transform the situation by using it as a tool for greater realization. We can receive the energies with gratitude, love, compassion and confidence that it is for the good of all involved. By holding this vision, we change the energy signature of the situation, resulting in a higher-vibration experience. There is never a need for defending ourselves from low-vibration energy, when we constantly focus on high-vibration energy patterns of joy and gratitude, because we are the transformers.

2.

Moving Into Ascension

Celebration of Ascension

Jesus was the master of his life, and he drew others into alignment with his vibration. He told us that his essence is our essence. Everything that he could do, we can do also. We need to take this seriously. I expect that soon we'll see lots of people walking on water. Many of us are very close to complete alignment with our higher Selves. In this perspective ascension can happen. We still have some DNA transformation needed to operate in a higher dimension, because our DNA has been disabled and is now in the process of rebuilding and transforming. All of this is observable in the areas where our natural vibratory frequency experiences anomalies in its wave patterns. These are our problems, all of which we have personally created for our experience. Now they must come into alignment with the energy of our heart. Once we are aligned with unconditional love and joy, we can know our true essence of Being, which vibrates through our heart. It expresses itself with high-frequency emotions in alignment with our heart energy. Here we experience compassion, forgiveness, deep understanding, unconditional love, peace and joy, if this is what we allow ourselves to experience.

If we can recognize ourselves as an eternal presence of self-awareness, we can live without limitations of any kind, because we have conscious access to the creativity of the quantum field. Our emotions radiate the energy that becomes our experience. We are constantly modulating energy patterns with our consciousness. As we become more perfectly aligned with our intuition and highest emotions, we become clearer and more expansive in our perspective. We can complete the regeneration of our bodies with our consciousness, and we can ascend whenever we're ready. We can live in both dimensions. We attract others who want to awaken to their true Selves. This is how humanity ascends. We can do what Jesus did. We can ascend.

Our Inner Guidance

Each of us has our own unique energy signature in the quantum field of all potentialities. Our vibrations attract other people and situations that are in resonance with our own. We do not need to know or feel anything more than we are given by our higher guidance in every moment. We always have feelings that urge us into higher vibrations, if this is what we want and are paying attention to and are expecting. If we are free of our old restrictions, there are no limits as to how high we can go. Once we begin in earnest expanding our conscious awareness from within our alignment with our heart energy, we are supported by the quantum field, which naturally brings everything we need to pursue our deepest passions in freedom and sovereignty. Fear of any kind becomes a distant memory. There is only unconditional love in every moment and in every being that we encounter, if that is what we expect and give our attention to.

In this state of being we are living in a realm of high-vibrational feelings and ideas.

The transition from ego consciousness to complete trust from within in every moment can be done with our feelings. The mind needs guidance. We can feel what to think. We can do this from an inspired state of being. We feel subtle urges that we know intuitively are the way we want to go in every area of life. In this way we free ourselves from the Matrix and begin living magically. Our expressions are expansive and inspiring. We are aligning ourselves with the divine Being that we really are.

The consequence of living this way raises our energy signature to a level that provides experiences that resonate with us. Everything in our former life must resonate or be left behind. This can be scary for the ego, whom we love compassionately as we bring her/him into alignment with our higher vibrations. Our confident belief in our inner guidance gives assurance for the ego to relax into peace and joy, just like the rest of our Being.

All of this is a stretch for awakening beings. We don't need to realize everything immediately. We just need to learn to be extremely sensitive to our subtle feelings in every moment without mental interference.

Living in Ascending Energy

As the Earth keeps rising in vibratory resonance, the historical dimension that we have all lived within is dissolving, and everyone who keeps trying to make things work the way they always have is finding that the old ways must be transformed. We can no longer live on a planet where people are destroying one another, competing for wealth and living in ignorance of the consequences of our intentions and actions. The frequencies of all beings must rise with the Earth's increasing resonance or become unstable and dissolve. With these changes, great chaos and destruction is inevitable, while the low-vibration dimension and all who identify with it struggle for their continuing dominance.

Along with our ascending planet, our Sun is increasing its brilliance, as it increases its vibratory frequency. We are being enveloped almost daily by massive influxes of gamma-ray photons, increasing our inner light and upgrading our DNA. We're being prepared for a leap in consciousness to a higher dimension of being. Our awareness is expanding to knowing in heartfelt ways the connections we share with one another and with all conscious life.

Quantum physics has shown us that we are not alone in any way. Along with everything and everyone who exists, we all share the Source of our being in the unified quantum field. Although we have successfully created a dream world of empirical separation in our awareness, this realm is now approaching terminal existence, and we are being released to realize our true Being. All of us will eventually awaken to our personal truth in Being, as the illusion continues to disappear.

We are not who we have believed ourselves to be, but deep within we know intuitively that we are playing a game of personal limitation. Since we have designed and committed ourselves to participate in this experience, we are the only ones who can release ourselves from this form of enslavement. We can cry out for money, health, freedom and romance from others, but such help will always be temporary. The only real solution is Self-realization. We must awaken through the highest vibratory feelings that come through our heart and let our awareness flow with the unconditional loving life force that fills our Being every moment. This requires our interest and intent to be open to this energy and embrace it with deep compassionate understanding for ourselves.

Living a Most Inspiring Life

Quantum physics has shown us that consciousness is universal, and that every particle and wave are consciously connected. In spiritual terms, that means unconditional love is universal, because intentionally recognizing that our constituent atoms and sub-atomic particles all communicate with each other constantly in the same consciousness, enables us to open our awareness to this realm. Reason informs us that every part of our body, including our emotions and mental processes, are brought into being in a conscious entanglement with the photons and other sub-atomic and atomic waves and particles comprising our presence as human beings. We participate in universal consciousness. Most of us are oblivious to this and do not believe it.

Transformation of our awareness is necessary to move intentionally beyond our current boundaries and limitations. We live in a spectrum of electromagnetic wave patterns comprising our personal energy signatures. To raise our vibrations, we can react to challenges in our lives with compassionate wisdom and understanding. We can also imagine high-vibration scenarios and feel emotionally present in these vibrations for periods of time. We open our awareness to greater consciousness, and we move into the realm of true compassion, joy and peace. This is a path to realizing our divine Self.

Once we can recognize our intuition arising from the energy of our heart, we become powerful in our radiance of divine love and joy. This has not been normal for humans, but as more of us raise our frequency with the will and intention to live in high vibrations, we change the vibratory frequency of humanity. This radiance is an invitation to others to awaken to our higher Being. This is how we make the conscious leap into a higher dimension of living on this planet. Once enough of us believe in the truth of who we are, humanity will transition to a higher dimension or spectrum of frequencies. In this we are being supported by the massive waves of gamma ray photons from the Central Sun, that are raising the resonant frequency of the Earth and humanity, as shown on the Shumann Resonance graph. We are in the midst of cosmic changes in frequencies, created out of the quantum field, which is the source of our Being.

Guidance for Humanity

We are being moved to live in deepest love, deeper and greater than we have ever imagined. We are being moved to be inspired in wonderful ways in all areas of life. We are being stimulated into joy and beauty beyond measure. It is flowing through us in the life force that constantly arises for us from within the quantum field. We are constantly coming into Being in the essence of the Creator of all and being expressed as ourselves with our personal energy signatures. We are fractals of divine consciousness, intimately connected in the flow of Creator energy.

As humans we have not been aware of any of this, unless we have begun to expand our consciousness beyond our human energy patterns. We have the ability to choose whatever we want to pay attention to in our experiences both empirical and imaginary. We have the ability to envision and feel whatever we choose. This is our creative ability, which we use constantly. We are the modulators of energy patterns in the quantum field. Most of us have not realized this, but we can open ourselves to greater love and wonder in the vacuum fluctuations of quantum energy. We can feel the quality of all of the energy patterns that we encounter and imagine. By choosing the quality of energy that we want to experience, we receive visions and feelings that attract the qualities we have chosen to align with. This is how we modulate the energy in our presence and create the situations that we experience.

By choosing to imagine being in unconditional love and joy, we are aligning with the flow of life force that we receive constantly in our consciousness. We can choose to be aware of this alignment and be moved to live in this quality of energy in the greatest love, wisdom and compassion for ourselves and everyone. We are transforming our lives into a higher dimension of Being and raising the energy signature frequency of humanity. We are aligning with the rising energy resonance of the Earth. We are expanding the light of our heart. In our radiance, we are emitting conscious photons in alignment with the energy of our Creator. We are the transformers of humanity's energetic awareness.

Quantum Conscious Being

How can we be our higher Self? What keeps us from being able to live in constant awareness of unconditional love or even to realize what it is? Our limitation is our complete focus on the vibrations we have been programmed to live in as humans. We have strong belief in the reality of the material world and the low-vibrational energy experienced in it. We believe we are mortal.

If we think about ourselves logically, in terms of quantum physics, it's impossible for us to be mortal, but we have to know who we are in our essence, beyond our ego consciousness, because the ego is mortal. We have designed it that way to make our experiences in this dimension as real as possible. Quantum physics has proved that consciousness is the basis of everything. There is nothing that is not part of universal consciousness, which is timeless and without space. It just is present cosmic creative self-awareness. It is continuously creative through us. We are fractals of cosmic creative Self-awareness. Humans have called this awareness God, the Creator of all. It is the awareness within us, our self-awareness. It is just present eternal self-awareness, and it can manifest anything we want to be or do, such as be a human being or a star or a galaxy.

We are the resonators of unlimited energy patterns. We can be creators of worlds and universes. Our consciousness works with vibrational frequencies of energy patterns. Our conscious state of being expresses itself as our personal energy signature. The frequency of our energy signature varies with our thoughts and emotions. A belief in our mortality is a false, low-frequency belief and keeps us from realizing our true Being. Belief in personal aging and mortality must be recognized as false by knowing our own self-awareness and the vibratory frequency of our thoughts and feelings. All of this knowing has no material representation. It is part of our self-conscious Being. And there is much more to expand into as we raise our vibrations.

We begin to awaken to our true Self by recognizing our knowing that we are alive and self-aware. This is our Being that expresses part of itself as human. We can consciously separate our awareness from our body. We do it in daydreams. Once we can make this change of focus intentionally, we can be confident in our eternal Being. We have expanded our self-awareness beyond our material manifestation. We are learning to work with energetics. As we learn to feel the promptings of joy and love from our eternal heart energy, we're able to live in those high-vibration feelings. We can intentionally come into resonance with our true Being, our higher Self.

Moving Ecstatically Through our Incarnation

The essence of all life is unconditional love, which few humans are aware of. We often write and speak of unconditional love, but what is it really? It is the constant conscious expression of the Creator into the vacuum fluctuation of the unified quantum field flowing out into all conscious entities in existence. It is the connection of all in the universal consciousness of the Creator. What we normally call love is an energy of attraction and compatibility. It is a desire for union. If we continue on the emotional continuum from here, we move toward the desire for union with the Light, the essence of the life force in every Being. When we recognize this Light, our lives become ecstatic in every way.

From the actions of the minutest waves/particles to the interaction of galaxies, everything in nature is a dance of perfect operations of life in high vibrational energy of unconditional love. Being infinite, it provides all that is needed to enjoy a high-quality life for every being. While incarnated with humanity here, we can naturally live as humans being well-cared-for on all levels of life. This is our natural state of Being beyond the human experience, when we are just Self-aware.

Being Self-aware for persons who are moving into high-vibrational Being, is a state of being clear mentally and emotionally. Just being aware can bring into focus the wonders of natural energy patterns into our being. Here is where we find the beginning of awareness of unconditional love.

At this point we can be aware of our personal essential Being. This is beyond the ability of the mind and ego to imagine. We know it through our emotions. In this focus of our Being, we find joy and abundance flowing through our lives. We are beginning to understand unconditional love as the essence of our Being as creators of infinite ability.

From the perspective of an enlightened Being, living among humanity becomes a compassionate and interesting life with the radiance of the One Consciousness that is the Creator of all.

Interacting with the Energies of Humanity

We face many severe challenges that are being imposed upon us by forces beyond our current ability to control. We are rebelling against detrimental contracts that we have been tricked into. We have been controlled by malevolent intelligence beyond our traditional capabilities. We have been forced into warfare against our will under threat of fines, imprisonment or death. This is the world of human experience that we now must transform beyond the ability of the dark force to affect.

The world that we have perceived as real is actually made of electromagnetic energy patterns that our consciousness interprets as the empirical world. Everything is energy waves and patterns that become empirical for us when we recognize them as such. Apart from our recognition, there is no empirical realm, only energy patterns. Every conscious entity lives within the energetic frequency patterns of whatever dimension it was created for, while being aware of universal consciousness that enlivens and creates everything. Although we've been living within a limited spectrum of the energy patterns of consciousness, we have the ability to penetrate the consciousness of our Creator. Our life force flows to us constantly from within the quantum field of all potentialities. This field is what physicists call the vacuum fluctuation, which is a constantly changing complex of every possible energetic pattern and frequency. It is an expression of universal consciousness.

We are fractals of infinite and eternal consciousness that have severely limited self-awareness in order to expand our consciousness into knowing what we feel when we're locked into a low-vibration environment. This human experience has given us wisdom that we could not have without knowing what this low-vibratory spectrum feels like. These are the vibrations that stimulate fear in us.

Without fear, we are not subject to the dark force, because we are naturally Beings of love and joy, and our high-frequency energy signatures cannot be affected by the low-frequency energy of anything having to do with fear. If we can focus on high-vibration scenarios, we can realize who we really are as pure, personalized conscious Beings with infinite abilities, all participating in universal consciousness, which flows with unconditional love in everything. We can transform the human energy spectrum.

Gaining Deeper Understanding and Wisdom

In our essential Being we are unlimited in our awareness and creative abilities, but we have agreed to participate in a very limited spectrum of creative expression and awareness. We chose a low-vibration dimension for our interaction with one another. We made it so compelling in capturing our awareness and locking us into a constant state of fear, that we have lost our sense of Being, but we can reacquaint ourselves with our complete Self.

If we feel attracted to high-vibration feelings of love and joy, we can transform our fixation in the low-vibration realm of fear. We can open our awareness to higher vibrations by paying attention to the energy of our heart and the flow of our life force. When we're learning how to do this, it is helpful to be in nature, especially in beautiful places with birdsong and fragrant surroundings, where we can relax completely and align ourselves with the energy of the Earth. The Earth is a multi—dimensional being, and as her vibratory spectrum rises, the lower vibrations of humanity are now out of alignment. They are becoming unstable, and those who dwell in that spectrum must rise into resonance or terminate.

Everyone is being drawn into a higher dimension of living. This is the natural flow of our life force, which is enhanced as we become more consciously expanded and transparent. Our Earth Spirit Gaia is drawing us higher in our awareness. She is regenerating and becoming more beautiful for us to perceive in a higher-vibration perspective. If we believe that Earth is just an unconscious thing, we cannot perceive the regeneration, which is in a higher-vibration spectrum, that is beginning to be expressed for all to know.

We are empowering this transition with our awareness of it and our emotional alignment. When we are aware of the natural flow of our life force within the universal consciousness, we are in alignment with the consciousness of Gaia. Through our attention, we add the flow of our life force to this expression of energy. It is all a reflection of our own recognition and realization. This is what we choose to think about and feel. We have the choice in every circumstance to express ourselves in the truth of our Being. By adjusting our perspective to come from our eternal awareness of compassion and love, we elevate the consciousness of humanity.

Modulating Our Vibrations

Humanity is preparing to awaken from the spectrum of vibrations that we have focused within for eons. Many have already gone beyond our traditional vibrational spectrum into both very deep and dark low-vibrations as well into very high, radiantly bright vibrations. We are all creating a vibrational level that we prefer to resonate with.

The Earth is creating a higher-resonating frequency pattern, which keeps rising. Humanity must adjust to this higher-vibratory spectrum or be unable to stay on this planet. We are all responding to these changing conditions in our own consciousness. If we choose to rise in resonance with Gaia, we can adopt the perspective of compassion and joy in every interaction. We can expect to see the light in everyone and to relate to that light. We can transform the vibrations of everything we encounter from fear into love and joy, because we recognize our expanding awareness. We can choose to be in joy and to interact with compassion and wisdom, even in low-vibration situations.

Chaos and destruction on many levels can happen around us, and we do not feel that energy, because we can direct our emotions into higher vibrations. Our personal experiences begin to come from the vibratory level of the life-stream flowing through our heart. To be successful in gaining a higher perspective and living it, we must be intent on it. We must act, guided by our intuitive knowing. This is where our attention must be receptive.

Whatever we do with our thoughts and emotions is a creative act as well as a momentary experience. By directing our feelings and awareness toward high-vibration perspectives in every moment, we are expanding into our eternal Self-consciousness. We can align with the unconditional love that streams into our Being continuously as the essence of our Creator. This is our process of awakening.

Our Return to Light

We were created in Light, and to Light we are returning. Incarnating into this limited human dimension has removed our memory of who we are, but we can regain the truth of our Being. Out true Being has not changed its vibratory spectrum during our lower-dimensional experience. Our reality is just awaiting our recognition. We are here to create from the energy of our heart. We know what that is because it's what we feel best about, the feelings that inspire us and stimulate our sense of gratitude.

If we can't feel these high-frequency emotions, we have dropped deeply into low-frequency vibrations, and need assurance and compassion from our heart energy. Here we can help one another energetically, when asked.

Our interactions naturally are from complete sense of presence of Being. Our intuition is clear and direct. We can feel the direction of our thoughts. With these feelings, we can transform low-vibratory patterns and guide our conscious mind to make high-vibratory decisions about focus and action. Every feeling and thought is important for informing us of the quality of energy we are facing and for creating the quality of energy we enjoy the most. We both become informed and become creative by receiving the energy we encounter. We receive information, and then we can react creatively with the quality of energy that we want to experience. We react with compassion and wisdom, regardless of the level of vibrations that we encounter. We move emotionally completely out of fear and into the spectrum of vibrations of love and joy. This is where life becomes miraculous, and our consciousness expands into greater knowing of everything.

By being in compassionate wisdom, we can transform any energy we encounter in our own experience. Everything that appears to be outside of our body is also within our consciousness, or we could not be aware of it. It is all a reflection of the quality of vibrations of our personal energy signature. The unified quantum field expresses experiences for us in every moment, regardless of what the former moment expressed. This is how miracles happen, when we react to low-vibrational encounters with high vibrations. It does not matter what appears before us. It's all energy patterns that we can modulate.

We can become aware of our true eternal Being and our infinite abilities. We are the ones who provide experiences of all kinds for our Creator, whose Being we are fractals of. We are Beings of unconditional love and limitless powers. In our nature we are masters of the modulation of energies. We're learning to be trustworthy by awareness of our intentions and feelings as we reside in high-vibrations as much as possible.

Expressing our Divinity

We are complex beings with significant abilities that we are hiding in the depths of our consciousness. We've designed our current self to be unaware of our true Being, until the time when we are ready to complete our foray into low-vibration environments. Many factors suggest that this is the time. The Shumann Resonance Graph, which measures the vibratory frequency and intensity of our living, conscious planet, is rising in amplitude, frequency and intensity, octave by octave. Historically measuring 7.83 cycles per second, it is moving up in four octaves to around 24 cycles per second. Physicists don't know how high it will go, but probably beyond 40. At such vibrations, all conscious life on Earth will need to be transformed in consciousness, in order to resonate with the vibratory frequency of the planet. No matter how low humanity may vibrate, we cannot overcome the energy patterns of the Earth, and we must align with them or go somewhere else.

In addition to the increasing resonant vibratory patterns of the Earth, we are now enveloped in a cloud of conscious photons, which are high-vibration sub-atomic entities. We are also receiving more gamma ray blasts of photons from the sun and beyond, which have the ability to change our DNA. This is beginning to transform our physical cellular structure into higher-dimensional, regenerated bodies. All of these things are indications that personal transformation and expanding consciousness is where we're going. We need to remember who we are, from before our incarnation into limited awareness.

Without our physical bodies, we are just a Self-aware presence of Being with unlimited creative abilities. We've created ourselves into our current situation, and now we can create ourselves out of it. Although some of us have been under technological control by dark entities, this arrangement cannot continue to exist, as the vibrations continue to rise.

We have absolute control over our mental and emotional processes and states of being. These are our creative tools. We can imagine wonderful scenarios and feel ourselves being in them, experiencing joyful and loving interactions. Through our attention, our mental and emotional processes vibrate at the frequencies that we feel and imagine. This vibratory pattern attracts resonant frequency patterns in the quantum field for us to experience. We can keep creating higher and higher frequency patterns that stimulate greater loving and joyous feelings, which bring expanding conscious awareness of our Selves, our eternal true Being.

Our Inner Transformation

We have accepted and aligned with the energy spectrum of humanity. It is a low-vibrational spectrum of experiences based in fear. Its basic assumption is the belief in our mortality. It is our identification as a physical embodiment with no way of knowing if there is life beyond the body. It ignores some obvious disproving facts that we know.

One is dreaming in full awareness of being in another realm. Day dreaming sometimes takes our conscious awareness into a parallel world, or one based on our past experiences. During these dreams, we are not aware of our body or physical environment. Our awareness is temporarily apart from the body. Deep meditation can result in traveling in full consciousness to wherever we are heart-felt to be attracted. By following the prompting of our intuition, we can use our imagination and our emotions to guide us toward higher-frequency visions and elevated emotions.

Unexpected ejections from the body, as happens from lethal accidents or attacks, and then returning to body consciousness leaves a memory of a greatly expanded sense of awareness. This completely disproves the assumption of termination of our awareness at the death of the body. It is an experience that anyone can have, and tens of thousands have. Many have recounted their experiences out of the body. Some of us have also developed the ability of astral and soul travel out of the body at will.

All of our experiences in the body are limited by our beliefs about ourselves. These false beliefs are disproved by our knowledge of immortality. Our eternal Being is not speculative. We have proved its truth. We only need to recognize it and realize what it means for us. And so begins the next part of our journey inward to knowing our true Self.

Finding Our Inner Light

We are taught by enlightened Ones that our divine Being is within us. The light of our Creator is within us. It is the source of all of the photons that we radiate through our aura, and the life force empowering our conscious energy signature. The challenge for awakening ones is how to access this awareness, if we are searching for it.

We have trained ourselves and formed strong beliefs about the boundaries of reality. We have become reliant on our physical senses and restrictive on our higher emotions and intuition. If we have a perspective of fear of any kind, true exploration of consciousness is not possible. We have been fearful of the unknown, including physical death and death or transformation of the ego. The ego can be released from its self-appointed role as sole guide in our human experience. The ego does not know higher guidance. It just uses its limited awareness to stay alive and live with minimal stress.

When significantly threatened, the ego becomes frantic and fills our awareness. We could easily lose our perspective and drop into fear. These fears are false, when understood from a perspective of love and compassion. Fear is a low-vibration energy, and it feels uncomfortable or worse.

To resolve all fears as they appear in our awareness, we can assume the perspective of eternal Being with infinite creative power. This is a major leap in consciousness to a high-vibration perspective. It feels confident, knowing that all of our vibrations are constantly attracting resonant energy patterns in the quantum field for our experiencing. If we can stay in high vibrations, our life experiences align with this energy. This requires our focus on the energy of our heart. This is where our life force flows through us in unconditional love out of the consciousness of the Creator. It is our inner light, which we can become aware of primarily through our feelings. It is the source of our joy and love and all high-vibratory emotions. We can feel this vibratory spectrum in moments of ecstatic joy and wonder. Those are wonderful feelings to stay with. They attract more of themselves.

If we're searching for our inner guidance, there are many ways of becoming aware of it. Our vigilance is needed as we go through life. It is all a metaphor, with many suggestions toward higher consciousness. It can be helpful to align with the energy vibrations of the Earth in natural, wild places. Inspiring music, advanced yoga, deep and rhythmic breathing, Tai Chi and many other techniques can be used to calm our ego and relieve it of its stress. Then we can become aware of our intuitive guidance.

From a higher perspective threats are not possible, because we know that we are eternal, Self-Realized Beings with unlimited abilities. We can live in a world that vibrates in a high-frequency spectrum of energy patterns, which we have a natural magnetic attraction to align with. This is the direction that nature is moving us toward.

Calming Our Ego with Higher Guidance

Our challenge in finding our inner divine guidance rests in making peace with our ego consciousness. We can sit quietly and breathe deeply and rhythmically, listening to inspiring music, observing every thought that passes before our awareness, questioning its reason for being and how it feels. Then let it go, as another arises, and we examine it. We are learning what is important to us on a deep level. Thoughts that seek our attention arise from somewhere within our consciousness. If we can be present in a perspective of acceptance and compassion, our true nature will arise for us to recognize. It is brilliant and wonderful to experience.

Once we begin to recognize the energy of greatest love arising within us, we can allow it to envelope us in its high-vibrational frequency, expanding our awareness beyond our personal dra-

Chapter 2. Moving Into Ascension

mas and encompassing the vastness of eternal Being. This does not usually happen all at once. It is a process that we can practice every day.

Eventually we can become aware of how the beliefs that we have been trained to hold keep us from deepest knowing of who we are. We've allowed ourselves to be programmed not to question them, but question them we must, if we want to be free spirits in unlimited Being. We have been held in patterns of existence that are uncomfortable and demeaning by our beliefs about ourselves. We have believed that we must be subject to the actions of others outside of our own persons for our well-being or our deprivation.

When we settle into the feeling of our deepest Being, apart from any condition that confronts us, we can recognize that we are pure consciousness. We begin to expand into inner knowing, beyond our experience in the empirical world. We can recognize our ability to feel higher vibratory energy and to know that we have a choice always to focus on any patterns of energy that we desire, regardless of our seeming outer experience. It is our choice that creates the quality of our experiences. Our choices can be guided by our ego consciousness, which is rooted in fear for survival, or the ego can be trained to wait for higher guidance that we can be aware of in our intuitive knowing. Following our intuition transforms our lives into experiences of love, beauty and abundance. These are our natural conditions, apart from our conditioned beliefs, which have kept us enslaved to low-vibration energies.

Two Paths for Humanity

While the vibratory energy of the Earth continues to rise, as measured by the Shumann Resonance graph, its effects upon humanity and all life forms that inhabit our planet are becoming felt and perceived by those who are seeking awareness. There is a clear division of life paths being offered. Rising frequencies in all areas of life and being stimulate higher-frequency feelings and thoughts in us. We can feel this most when immersing ourselves in places of natural beauty undisturbed by humanity as much as possible. Old-growth forests, deep canyons with wild rivers, waterfalls, high mountains and tropical seas all have a powerful radiance of Gaia's spiritual presence. Here we can more easily align with the energy of our loving planetary Being. We are being carried into lives of joy and beauty beyond the low vibratory influences of the chaotic and tyrannical world of humanity's current experience.

The energy of Gaia is increasingly brilliant and filled with vitality and beauty. Gaia is regenerating her physical body. It has already happened in the quantum field, and is beginning to manifest in the material realm. We can contribute to this manifestation in our personal experience by intentionally maintaining a perspective in the spectrum of elevated vibrations of compassion, serenity, delight, love and abundance. As we increase our mental and emotional vibrations for sustained periods of time, we attract fewer encounters with low-vibration experiences, until we become invisible to their intrusions. This is how we transform our lives to align with the rising resonance of the Earth.

Humanity cannot maintain a perspective that is out of alignment with the vibrations of the Earth. Gaia is a very high-vibration Being and is much more powerful, requiring all discordant and anomalous frequencies in her presence to come into alignment or become unstable and dissolve into unmanifested energy in the quantum field.

As Gaia continues to rise in resonant frequencies, the persons who are refusing to come into alignment, because they're attached to low-vibration, fear-inducing situations and beliefs, are becoming unstable mentally and emotionally. They recognize their approaching mortality and are desperate and dangerous. Most of the world leaders fall into this group.

We do not need to participate in low-frequency situations. Because everything in our experience is a manifestation of energetic frequencies, the way to change our situation is to change the vibrations in our own energy field. Our natural essence of Being has unlimited creative power for our experiences, but our false beliefs about ourselves keep us from realizing this. All of these beliefs arise as a result of our fear of suffering and death. In the spectrum of the rising vibrations of Earth, there is no suffering, and we recognize our eternal self-aware presence of Being.

Our Etheric Being

We live in an environment where nearly all of the energies that envelop us are unknown to us. Even the essence of our true Being is hidden from our awareness. As infants we telepathically learn how to think and perceive our world as humans, and then most of us learn that we're not telepathic and that the only reality is the empirical world. We are subjected to influences of energies beyond our recognition and have unwittingly allowed ourselves to be immersed into low-vibration lives of stress, suffering and fear. We have learned hatred, jealousy, persecution, shame and even murder, and we have projected these feelings and experiences upon others. It is all pretense and exists only in our own consciousness.

Quantum physicists have discovered that the entire cosmos is filled with electromagnetic and etheric energy beyond our human awareness and contained in universal consciousness that is the essence of everything. The energies that we recognize, focus on and feel become part of our experience. We have absolute control over what we allow to influence us. Our consciousness expresses itself as patterns of energetic waves which interact with all of the energy that we may not be aware of, because of the belief systems that we have acquired.

Each of us incarnated into a situation that is bound by the beliefs of everyone around us. The qualities of our lives are a result of those beliefs about the nature of our reality and the limits of our awareness. Although our limiting beliefs are well-developed and live deep in our consciousness, we are not required to maintain them. We can recognize them for what they are. Our belief of mortality, for instance, is false. We know this from all of us who have taken our conscious awareness out of our physical bodies and out of the empirical world into the realm of pure unmanifested energy. Here we can recognize anything that vibrates within our own energy spectrum. In our true Being we are free and sovereign eternal entities endowed with the universal consciousness of the One Creator, able to experience anything we desire.

We have incarnated as humans on this planet now, because of the turning of the ages, for our guidance into the new world of higher frequencies of greater love and joy. As we penetrate our consciousness beyond our limiting beliefs, we become the Creators of the regenerated human experience.

Our Divine Presence

We are much more than human beings, but our true identity has been hidden from us upon our incarnation, so that we can have a valid human experience. Now it is time to remember who we really are as Beings of light and love and infinite creative ability. It is no longer necessary to subject ourselves to the chaos and tribulation happening all around us, because the Earth is ascending into a higher spectrum of energetic vibrations, and we can transform along with her.

We can focus our attention on the life force that flows through our heart with full conscious awareness and unconditional love. We can elevate our conscious vibrations in alignment with the rising frequencies of the Earth. We can resolve all of the low-frequency beliefs we've accepted about ourselves, forgive ourselves for being stuck in fear for our lives, and open our awareness to the awesomeness of our eternal presence of Being. We can recognize the impermanence of this world and everything we've held onto, thinking that we need to depend upon money and physical nourishment for our bodies, when actually it is our conscious focus that attracts all of these things into our experience or keeps them away from us.

We are Beings beyond time and space with unlimited creative ability. We are fractals of the One Universally Conscious Being Who is the essence of everything—all energy, all beings, all that exists. We have compartmentalized our consciousness to incarnate on this planet, and we have the ability to realize our true essence. We can know that we encompass the universe and beyond. We are unlimited in our Being in every respect, and we can inhabit our bodies in the truth of our greater Being. If we choose, we can maintain our bodies in perfect health for thousands of years. We can move instantaneously in our presence of Being to other galaxies. It is all dependent upon our personal beliefs about ourselves and the transparency of our focus within universal consciousness.

When we maintain our essence in the high vibrations of compassion, love, serenity and abundance, we are beyond the reach and influence of those stuck in low vibrations—the dark ones and shadow beings, who seek to use our life force for their nefarious intentions. As the resonance of the Earth rises, they will disappear. We become exempt from their attention, because we can live in a different reality, a higher dimension of vibrations. This can be our new life situation. It is within the realization of all of us, if we choose it and train ourselves to believe it. This is a process for us. For most of us, it cannot happen quickly, but we can attain it with diligent application of our intention.

Awakening as an Intentional Process

In awakening from the human hypnotic trance of living in the Matrix, most of us go through a process that begins with our intention to be fully conscious. We can practice being present in our awareness, not thinking about anything, just being aware and focusing on awareness. We can practice this until it is our usual state of Being. At some point we become aware of subtle promptings in our intuitive knowing. We feel these as heightened vitality and stirrings of love and joy and expressive power. We automatically know what to do in every moment. We are not aware of these promptings, unless we are open to them in a state of serenity and gratitude. The entire process requires our intention to be aware in our presence of Being. This is a high-vibration state of Being. Its frequency is beyond fear and only in the presence of joy and love. Here is truth and innocence of spirit.

Sometimes we can make this transformation instantly, through experiences like being struck by lightning or having a death experience from which we return to our embodiment. In a flash we know who we are beyond the embodied expression of ourselves. Even this kind of experience is only a beginning part of the process, because we are much larger than we can imagine. This again requires our intention to rise into even higher vibrations. Once we begin this process, we fall in love with it, because it's so wonderful. Every step brings more freedom and intentional creative expression. Our ability to express the energy that flows through our heart attracts energy patterns that align with these vibrations. Our quality of life arises from these high-frequency energy patterns. Life becomes much more fun and meaningful.

We may be living in situations that don't change in quality, life continues to present challenging situations to us, and we have a lot of mixed feelings about life. Perhaps we've been meditating and have achieved a state of serenity, but this is transitory. Our life in the Matrix keeps demanding our attention, and attempting to keep us in fear and guilt and shame.

If we want to expand our consciousness, we need to withdraw from alignment with the low vibrations around us. We must realize that we have the freedom to be however we want in every moment. We are not required to think or feel within the spectrum of energy that confronts us. We can deal with it from a higher vibratory pattern by intentionally opening our awareness to our presence of Being. Just being aware, until we know intuitively what to say or do. We don't necessarily gain this ability quickly. It requires intentional practice, as is true for the entire process of enlightenment, but we know when we're making progress, and every higher vibrational pattern that we experience feels fulfilling.

Transforming Our Lives

In becoming aware of our serene presence of Being, just being in awareness, we can receive the inner guidance that most benefits us in every moment. We do not need guidance from anyone outside of ourselves. We have a direct connection to the consciousness of the Creator of universes. We receive our conscious, unconditionally loving life force flowing through our heart constantly. The only way that we can be in need of anything or fearful for our lives is by our own self-imposed limitations. We do not need anything from anyone outside of ourselves, because our essence is being infinitely powerful Creator of everything. We share the consciousness of our Creator. It is who we are.

As humans on this planet, we have incarnated without knowing who we are. We have been taught since infancy that we are limited in every way and are dependent upon others for the fulfillment of all of our needs. Our religions have taught us to pray to a God that is separate from us. We have been forced to work for others in order to survive and to pay taxes to our governments, which then deprive us of freedom. All of this is a great contrast to our true Being and creates a difficult situation for anyone who is on the inner journey to awakening.

Those of us who are on a spiritual path have already decided to awaken, and now we face many obstacles to our confidence and experience. We wonder if we really are the creators of our lives, because we still face all the complexities and challenges of human life, both inner and outer.

Quantum physicists and quantum psychologists can offer some substance to knowing about ourselves. It has been proved that when we intentionally observe and recognize sub-atomic electromagnetic energy patterns, these wave patterns immediately become the material particles that we anticipate. There is an interaction between human recognition and energetic waves in the quantum field. There is no doubt about this. Science has discovered the link between intentional observation and energy modulation that results in materialization. This implies that there is intelligence involved in this process. Science has not yet discovered the emotional process of creation, but once we have opened ourselves to our intuitive knowing, we realize that our emotions are intimately involved.

Our bodies are not wired to be able to perceive sub-atomic particles or feel their presence. We have designed technology to observe them, but we don't have emotional technology to encounter and report to us the feelings of photons or electrons. We need masses of sub-atomic particles to reach our human threshold of recognition.

The implications are that our creative ability consists of thought and emotion. Our thoughts determine the patterns of vibrations that we create, and our emotions provide the frequency and intensity, the wave amplitude, of how we feel. In our creations, there is no difference between what we imagine and what we physically experience. As long as we align with the frequency wave patterns of our current situation in life, our circumstances will not change. We just keep recreating them. Only when we understand them from a high-frequency perspective, can we believe that we have truly creative ability. Then we can transform our lives.

Our Life as a Metaphor

Is it possible that our physical life is not what we believe it is? To realize the answer, we need to pay attention in a perspective of curiosity and wonder to everything that is happening in our experience and all around us. We need to be aware of the energetic frequencies that we feel and the images that come to mind, because it is our recognition of energy patterns in the quantum field that creates our experiences.

Every thing we perceive and feel is symbolically significant. Everything has an energy signature that is either gaining frequency or losing it. The Earth is going through a transformation in resonant frequency, as Gaia's consciousness expands into higher octaves of frequencies. In the higher dimension or spectrum of vibrations, there is no incursion of lower-frequency energy. Gaia must resolve all low-frequency energy patterns held in vibration by humanity. They will come into alignment with Gaia's ascension, or they will destabilize and disappear into another dimension. Gaia is clearing her consciousness, as she begins to regenerate. The human energy signature must transform from fear to love, as the energy of Gaia will no longer tolerate fear-based, low-frequency thoughts and emotions.

Gaia is rising into a higher dimension and will no longer be available in the old dimension for humanity. All who wish to stay in the current world during this incredible transformation will need to be given a different, but equivalent home and be able to carry on as they wish. The transformation will not affect their quality of life. The Creator deems that everyone proceed at their own pace. This is a completely free-will planet. We must give our permission for any violation of our person, and we can intentionally withdraw that permission at any time by raising the vibrations of our focus, which can now be based in the frequencies of love.

Quantum physicists have discovered that consciousness is the source of everything. That includes every aspect of each of us. We are the expressions of consciousness. Our consciousness, empowered by the life force flowing into us personally from the universal consciousness, constantly creates our bodies. If we are not limited by beliefs about our boundaries of consciousness, we can intentionally change anything about our bodies. Our bodies are energetic expressions that operate electrically and have a magnetic emotional presence. We control the quality of experience of every part of our bodies with the energy frequency level of our emotions and thoughts. As we become more loving and joyful, we begin to regenerate our bodies, just like Gaia.

The empirical world that stimulates all of our senses exists as energy patterns that our consciousness interprets as solid material. This is the metaphor. Our reality is our interpretation of energy patterns. Our story may be written beyond time and space, but we get to choose the quality of our lives in every moment eternally.

Humanity's Only Future Timeline

The awakening of humanity from the hypnotic trance of eons is happening now. All possible timelines for our experiences have converged, and there is only one possible direction ahead. The increasing flow of high-frequency divine energy is carrying us all into a higher dimension of vibrations.

As the resonant frequency of the Earth rises by leaps of octaves, all life forms on the planet are coming into alignment. This is why it's important to spend time in nature, absorbing and aligning with the natural vibrations and soaking up the Sun, whose transmissions are bringing us large quantities of gamma ray photons, making everything brighter and awakening our DNA into higher vibrations.

We live in the plasma energy flow of the cosmos and have the ability to recognize energy patterns through experiential and imaginary realizations. By constantly choosing to be in heart-centered situations, we elevate the quality of our energy signatures, attracting experiences that align with our vibrational spectrum.

We are being prompted to be more heart-centered in our perspective. If we imagine and feel ourselves being in wonderful situations vibrating in the spectrum of love, gratitude and compassion, we create the energy patterns around us in resonance. Fear becomes non-existent, once we know our personal eternal Being. This is what we are awakening to, as we become aware of our limitless nature. We participate in universal consciousness, out of which everything is constantly created moment to moment.

Our true Being is just present Self-awareness with unlimited abilities by virtue of our mental and emotional energy modulations. As we eliminate our false beliefs about ourselves, we can expand our awareness without limits into full realization of our true Being.

Moving Out of Suffering and into Fulfillment

Complete fulfillment for us happens when we focus on the energy of the heart of our Being, represented through our physical heart. The purpose of our heart is to enliven us and inspire us with deepest knowing, and it does this regardless of what we do to it. This is our clue in knowing how to interact with threatening situations and people. They are all limited by the low vibrations of fear. We have been in this spectrum of energy during most of our earthly lives, but we are not required to align with it. We can challenge it by aligning ourselves with the love vibration. We can understand every situation and person from the perspective of compassionate wisdom. When we do this, we elevate our energy signature out of the low-vibration spectrum of fear-based experiences and into a higher-frequency spectrum of vibrations. In this level of energy, experiences happen differently, because our emotional and mental attractions are generating an electromagnetic polarity that attracts wave patterns that come into alignment with us.

In conscious alignment with love-based, high-frequency energies, we can face any assault on our lower-vibrational human. While we remain in high-vibratory thoughts and feelings, we are not available for assaults, because we vibrate at a higher frequency of vitality. The human threat transforms into an experience of love, or it disappears into another dimension beyond our personal experience. This seems like magic, but it's just a different spectrum of energy beyond our present perception. Because our present awareness is limited by our beliefs about ourselves, in order to expand our consciousness, we must resolve all of our limiting beliefs, which do not allow for our eternal Being.

Our bodies age, not because it's natural, but because we believe that they age. We have taught our innate being to gradually withdraw life force from the body. We can reverse this by changing our belief, at the deepest level of our consciousness by intentionally calling up feelings of deepest love for our true Self, our own expanded conscious Self, who lives in eternal unconditional love and joy. Once we transform the fear of termination of our consciousness into knowing our eternal Being, we can begin to realize our multidimensional essence living beyond time and space. From our greater Self, we are projecting our human presence within a compartmentalized consciousness in order to experience intensely what would otherwise be impossible.

If we stop giving our life force to anything that limits us, by aligning our attention with the vibration of truth, we can clear our consciousness of all fear-based beliefs and limitations. For this we can be forgiving of everyone and all situations, accepting of all, and loving of all for all of the energies we have experienced. We have developed deep compassion and greater love as a result. We can resolve all of this and open ourselves to higher-vibrational visions and feelings. Life becomes wonderful in every way, because the higher dimension of experience that manifests for us is based in the love vibration.

Inter-Dimensional Beliefs

On the path of awakening to our true Selves, we must confront all of our training and personal beliefs. One belief that we have accepted is that we are mortal. From the spectrum of frequencies that we share with each other in the empirical world, this is real for our bodies. Upon leaving our bodies, we remember our expansiveness in other dimensions, and we become aware of our essential Being, our present awareness. We can recognize our innate creative ability, which we receive from our Creator, and we can learn to express the energy of the heart of our constantly-creating Being. This raises the vibratory resonance of our energy signatures into alignment with our natural state of Being. In this spectrum of energy, fear has disappeared, and everything is in the spectrum of love.

Once we know that we are timeless in our Being, all beliefs vibrating in resonance with fear become unstable and disappear. Our challenge is to align our entire selves with knowing immortality and to withdraw our emotional involvement from believing we are mortal. We can convince our mind from the written accounts of many who left their bodies and returned, but emotionally we may feel a draw toward our old beliefs. Without personally having an out-of-body experience, we can trust our intuition. This is what we know in our deepest Being.

Our intuitive knowing comes from within the energy of our heart. It has only fulfilled desires and conveys the life force of the Creator to the essence of our Being. Any beliefs out of alignment with the vibratory resonance of the Creator disappear. The premise on which belief in mortality can exist is that there is only one dimension, that of empirical stimulation. We can choose to accept other dimensions that we dream in and enter in deep meditation, as well as those we create in our imagination.

By intending to remember our multidimensional nature, we can focus our awareness in any dimension upon a moment of present consciousness, a moment of now that continues being always now. Our awareness includes all potential energy patterns in the quantum field. We feel every one that we focus on, and we attract those that resonate with us. We can open ourselves to intuitive guidance from within, apart from any influence we encounter or how challenging it may be for the ego.

Divine love is the only reality allowing everything to exist. It is the life-giving essence of Being constantly flowing from the Creator and enlivening all conscious entities. The world we are creating with this energy encompasses all life-giving energies. All low-vibration energies and the entities aligned with them are disappearing into a dimension that resonates with them, while we transform our conscious experiences into alignment with unconditional love.

Our Destiny as Creators

Quantum physicists have discovered that our world is created and dissolved trillions of times each second. It happens so fast that we are not aware of it, but what we are aware of is the flow of energy patterns that we recognize and accept as reality. This is the process of materialization. It is a psychic phenomenon, and we are the creators, who modulate the energy patterns that we recognize in our own consciousness. Nothing exists in the empirical world, except what we observe, recognize and accept as reality. Everything else is electromagnetic energy in infinite

Chapter 2. Moving Into Ascension

patterns of frequencies in the universal consciousness expressed in the quantum field of all potentialities.

Because we are part of the universal consciousness of the Being that constantly creates everything, we are constantly created as personal expressions of the Creator. We have chosen to accept limiting beliefs about ourselves so that we could experience a realm that would be impossible for us to truly know, if we were in our natural state of expansive consciousness. We are doing this to deepen our understanding of love and compassion by living in a world where this perspective is largely absent, so that we could know what it is like to live in fear for our lives, to suffer, to be greedy, judgmental and to desire to control others for our benefit. All of this would be impossible for us in our naturally created state of Being with our unconditional love and unlimited abilities of creation.

We are here to create experiences for ourselves and to learn what we truly value, appreciate and are thankful for. Currently humanity is experiencing much chaotic and destructive energy, because we are realizing the truth of what has happened on our planet and are clearing out all of the energy patterns that have been destructive of life. We have to face them all, bring them into the light, refuse to participate in fearful energy patterns any longer, forgive everyone involved and open our hearts to the unconditional love and joy of our eternal Being. We are the Creators, and we have the innate ability to align ourselves with life-enhancing energy patterns that will renew ourselves, our society and our planet.

As humanity awakens to the truth of our situation, we have a choice of continuing our current way of being or expand our consciousness to our natural state of living in abundance, kindness and beauty, realizing and utilizing our unlimited creative abilities as extensions of the Prime Creator. We do not need anything outside of our own Being for fulfillment and deepest love.

Facing the End Times

We are experiencing a planetary emergency unlike anything we've faced before. It's not about climate or earth changes. We are staring into the face of pure evil that is attempting to terminate human life on this planet. The ruling elite have led us into a realm of fear and destruction in every conceivable level of low-vibration energy. This is the finale of our descent into fear. As we and our neighbors become so distraught with lockdowns, masks, forced distancing from each other, starvation and forced injections of poisons and lethal parasites, we're finding it difficult to maintain faith that the light is getting brighter and the energy that envelopes us is rising in frequency. Let's take a few deep breaths and reevaluate our situation.

We incarnated here to experience feelings and thoughts that would not be possible for us in the higher dimensions of Creator energy, which is where we were created to be. We arise from a conscious Being that is unconditionally loving and expansive, and we have the ability to choose to experience whatever we desire. We wanted a challenge that would deepen our understanding of the qualities of life. We wanted to know what we could not know in the higher dimensions. We wanted to develop deeper love and compassion than we had been capable of. Our solution was to create the spectrum of energetic patterns and frequencies that we are now experiencing. Our stay here would not be complete until we came into the presence of the darkest, lowest-vibration energies that can exist. This is our final experience of living in the lowest vibrations of fear.

We now have the opportunity to use abilities that we have not recognized in ourselves during this incarnation. As we become motivated to transform our situation, we begin to ask for help from higher powers, not yet realizing that we hold within us the higher powers. Because of our free will, our lives cannot be interfered with by divine Beings. They can assist us, but only to the extent of directing us to our true Being. This is cosmic law. Our universe is constantly created moment to moment within these patterns of energy. We must become the masters of our own lives. We have participated in the creation of our current circumstances, and we must create our way back to higher dimensional experience.

We must awaken from the hypnotic trance that we've been experiencing as life on Earth and realize our true capabilities and the essence of our presence of Being. We are fractals of the Being that expresses Itself as the universal consciousness that constantly creates everything. We are not mere humans. We are the Creators of worlds. We are the modulators of all energy patterns that we recognize. By focusing our attention and feelings in scenarios filled with love, joy and fulfillment in every way that we can imagine, we can change the energetic environment that we experience. This is how we can encounter any low-vibration experience and any beings of evil intent. By choosing to hold a perspective of kindness and compassion, we transform all low vibrations into higher frequencies of beauty and peace. In this way we are creating our New World in a higher dimension. We are experiencing the energy of Ascension, and it all happens within our own expanding consciousness.

Our Call to Ascension

Our thoughts and feelings are like prayers that radiate out into the quantum field and bring back experiences in the level of vibration that we feel as we focus our attention. Because we live in a quantum field of energy, the most important elements for us to realize are that we are continuously creating our experiences; the vibratory frequencies of the energy patterns in our minds and emotions create a constantly-changing energy signature. Its radiance modulates the energies that we encounter and think about to create for us experiences that align with our vibrations. These are our creative abilities. When we are fully conscious in our true Self, we have complete control over our minds and emotions, because we are focused on high-vibration feelings and visions. This means living in the energetic spectrum of joy, abundance, gratitude and love. This realm is our natural state of Being, and it is what we are being drawn toward by the rising resonant frequencies of the Earth and our galaxy.

It is from this natural state of Being that we decided to enlarge our awareness and our ability to love. Since we could easily realize more love, we chose to deepen our love by experiencing its opposite in the dense low-vibration fear-based energy patterns held in place by the awareness of all of humanity. Then we also created experiences of servitude and slavery, and sometimes torture, for eons. We've endured the entire spectrum of fear-based, low-vibration energy, and now we can decide to return to our full consciousness.

The first step can be gratitude for our lives and everything we experience, even if this appears to contradict our current living conditions. Reality begins within us, not in any energy outside of us. There actually is no outside, because everything constantly arises out of universal consciousness, which we also arise from. We are all the same Being, and we all have access to universal consciousness and the unconditional love that connects us all.

Each of us contributes experiences to the universal consciousness of the Creator Being. Those experiencing the most difficult lives are to be commended for being brave enough to contribute to our knowing those experiences. Those experiences stimulate a greater deepening of our compassion, which is our purpose as seen from a high-vibration perspective.

To live in the spectrum of love, abundance and complete freedom, we must imagine that it's possible. If we don't thoroughly believe it, our consciousness becomes conflicted, and we create chaos in our experiences. Clarity is necessary for the best resonance. As we become able to imagine a life of miracles and wonders, we can begin to modulate energy patterns that we recognize as wonderful. The longer we can recognize high-frequency scenarios, the more compatible experiences we create for ourselves and everyone around us, as we energetically walk into a new dimension of living in alignment with the energy at the heart of our Being.

3.

Working with the Law of Attraction

Attracting Our Best Friends and Lovers

We are born with a destiny that we planned prior to incarnation. There are certain experiences that we wanted to have either for our enjoyment or to balance our karma. It all relates to our quality of life, and nothing happens by chance. The only variable is our personal evolution and expansion. This is determined by our free will in how we handle all of life's situations.

We always have the choice of living by our imagination and emotions in creative ways that feel expansive and loving. We can feel complete in our own Presence. We are attracted to, and we attract beings of similar vibration to our own. Once we can successfully resolve any anomalous energy that arises by aligning the energy with our higher resonance, we are no longer subject to our natal destiny and our inherited karma. We are free to be ourselves in our natural state in alignment with the natural rhythms of the cosmos and the quantum field. These are high-vibration energies of unconditional love and universal consciousness. In this state we are unlimited modulators of all energies that we encounter. We are the creators of our lives and the masters of our experiences.

In our natural desire to live with wonderful people, we naturally attract our soul mates and twin flames who are vibrating in similarly resonant patterns. Because our emotions are magnetic, our attractions are natural, and we can enhance them with our intentions. By creating imaginary scenarios with high-vibration emotions, we raise the vibrations of our energy signatures. By imagining living in the presence of our soul mates, we are creating that kind of experience. We have to do this until we absolutely believe that it's true. Our belief makes it real, just like Jesus said. Then our soul mates enter our experience.

Inspiring Our Lives

We can live in wonderful energy that uplifts us into great joy and ecstasy. Nothing in our lives happens by accident. It is all energetically connected in patterns of resonance. We don't necessarily create the images we have in mind, but we create the quality of what we feel in our experiences and imaginary scenarios. This is the frequency pattern in any given moment in our energy signature.

It is necessary to resolve all low-frequency vibrations that we're still unknowingly holding onto in our subconscious. They act as anomalous vibrations in our personal energetics, which will attract other low-frequency experiences. We can recognize them whenever they arise and cause us to feel or do something that's not quite right for us. There's always fear trying to hide as something else to avoid recognition. It never feels good or right. This is when that part of our consciousness needs compassion and embracing love to bring it into alignment with higher vibrations. Our true heart is unconditionally loving. It lives to give us life, regardless of what we do to it. This is divine energy of very high vibrations.

When we can be compassionate and loving, we can observe ourselves from a perspective of neutrality. There need be no blame or judgment about anything, regardless of how low its vibrations are. What matters is our perspective and energetic alignment. If we can maintain elevating feelings in every situation, we can transform our lives.

Our consciousness flows into us through our true heart energy constantly out of the quantum field. As we raise our vibrations intentionally, we create more loving and joyous experiences. Our emotions are magnetic in attracting resonating energetic patterns. This is where our experiences come from. Our energy signature radiates conscious biophotons into the quantum field of all potentialities and attracts scenarios that resonate. These situations are presented to us for our experience. High vibrations in our heart and mind create high-vibration experiences in all dimensions. This is the natural energy of our Being, as we are participating in the consciousness of our Creator.

Loving Beyond the Ego

For most of our lives we've lived under the impression that, although we can initiate projects and activities, most of our experiences result from influences outside of our ourselves. We believe that we are subject to conditions and situations imposed upon us by others outside of our own consciousness. We have developed thought patterns and emotions that we live within to maintain order and security in our lives. We have developed fear of the unknown, because we believe that we can be threatened or intimidated by powers greater than we can cope with, and so we live within our own psychological confinement.

We've learned that we are separate beings with our own vitality produced within our biological functions. This limits us in our expressions and creations.

At some point, however, something happens to expand our awareness, and we begin to realize that everything is happening within our own consciousness. Our entire lives are reflections of our own perspectives and expressions. We are the creators of our reality through the focus of our attention and our emotional state of being within the confines of our beliefs about who we are. One of those beliefs is our ability to love. While we believe that we are distinct, limited beings, we are limited in our ability to love. We set the conditions that make it possible for us to feel and express love.

Love is our attraction and connection to aspects of ourselves that we recognize as other beings and things. When we expand our awareness beyond our accepted limitations, we expand our ability to love. From the experiments of quantum physicists, we can infer that consciousness is universal. It is not limited to any entity, but is shared by all. We have learned to limit our conscious awareness to a spectrum of energetic expressions that we share with humanity in a realm of experiences that allows us to feel isolated and individuated. We can continue to live in this dimension as long as we choose to.

If we want to expand our awareness out to universal consciousness, we must raise the vibrations of our energetic presence by recognizing ourselves in a realm of natural joy and beauty. We can do this by intentionally aligning ourselves with higher-vibrational imaginings that gradually come into our experiences through the attraction of resonant energy patterns of situations, provided our beliefs allow us into this realm. All our psychological limitations and emotional knots must be resolved.

Once we begin to recognize our creative ability, we become confident in our personal sovereignty. This is when we can recognize the energy signature of unconditional love. It is the same vibrational spectrum as the energy of eternal Self-realization. Through our heart-felt intuitive ability to realize it and feel its vibrations, we can align with this energy spectrum. This alignment opens us to the experience of unconditional love, which constantly flows through us with our life force from the unified quantum field of the universal consciousness of the Creator of all. Here we can just be our Selves, creating what we want to feel and imagine for ourselves and all of humanity and beyond.

The Energetic Path to Infinite Love

The quality of our lives depends upon our vibratory spectrum, the frequency of our energy signature in every moment. Energy can change dramatically instantaneously, although changes usually happen gradually. For us, it all depends upon what we believe is possible. Energetically every moment holds infinite possibilities, but only the ones within our belief structure will materialize in our experience. We won't recognize anything else.

The quality of our state of being creates experiences that will stimulate the feelings we will experience. In our emotional being, it doesn't matter what we are facing except for how it feels. That feeling tells us the vibratory quality of the energy that is present. We can either align with that vibration or let it pass through our consciousness like radio waves on a channel that we're not tuned into. In this perspective we can be in our own emotional level while being aware of outer situations. These we allow to be as they are without judgment or desire, so that we may interact out of our perspective of joy and compassion.

We can align ourselves with the conscious flow of life force from the Being of the Creator. We know this energy by how it feels. It always feels good and can be intensified to be as ecstatic as we can handle. It is the constant flow of unconditional love into and all around us everywhere. It can be felt by those who desire to be aware of it, and it opens and expands our awareness, always by our choice.

We can become seekers of higher vibrations of being. This means higher frequencies of thoughts and imaginings and elevated emotions. If we can create wonderful scenarios in our imagination and align with high vibratory feelings, so that our entire energetic being resonates at high frequencies, we naturally radiate our vibratory pattern into the quantum field in every moment for the creation of the high quality of our experiences. Our sense of being becomes unlimited. We begin to live with the guidance of intuition, which comes to us personally in our life force from the consciousness of the Creator, and is felt emotionally and translated by our intuition for us to understand and know. This is how we know all of life's secrets and everything we need to be aware of.

The Reason and Purpose for Unrequited Love

There are many popular songs about heartache and loss in love relationships and much drama in society because of it. Many people think that we must protect ourselves from vulnerability in love, especially if we have suffered from the abuse of someone that we felt a deep connection with. What is really going on with this?

From the perspective of quantum energetics, we are all arising from the same universal consciousness, living in the same flow of life force of our Creator and enjoying absolute personal freedom of choice in all areas of expression. We have gone through painful experiences, which we created by allowing ourselves to focus on these experiences while suffering in them. We enhanced our creation of hurtful situations by fearing them before they happened. In order to experience hurtful thoughts and feelings and physical pain, we must align ourselves with the frequency spectrum of their vibrations. All of these experiences are prompting us to become aware of our creative power.

Once we recognize what we're capable of, we can take back the control of our awareness. There is no requirement to continue to invigorate any low-frequency situation. If we align ourselves with only high-frequency thoughts and feelings, we will not attract low-frequency experiences. We are capable of becoming exempt from personal attacks and difficult dramas by knowing in our expanding awareness who and what we actually are, and the abilities we have as creators, beyond the material world, but influencing it with conscious control of our vibrations.

On a more intimate level, we can split our essence and be in more than one place at the same time. In spiritual terms, we can express ourselves as twin flames, beings of the same light, emitting identical photons, but having separate experiences and making separate personal expressions and decisions, and ultimately coming back together. The timing for this attraction is now, as we are training ourselves to be completely free of attachments and intending to realize our true Being.

We exist in a sea of plasma consisting of infinitely diverse energy patterns that we can focus on and give our life force to. Through the focus of our attention, we create the radiance of our energy signature, vibrating in the quantum field, attracting and repelling energy patterns that either resonate with us or do not. Those that resonate will arise out of the quantum field as experiences. We can choose to be fulfilled and invulnerable in our eternal Being.

Choosing Our Destiny

Some of us live in life-threatening situations. Some of us suffer silently. Some of us fight angrily against oppressors. Life on this planet has many challenges and low-vibration experiences. What is the purpose for all of this, and why are we involved in so much struggle and suffering? Most of us believe that destiny provides our experiences, and that we must react to them in the ways that our ego consciousness needs to survive.

All of our experiences operate by the law of attraction. By focusing on our problems, we continue to create them. By fighting against our enemies, we enable enemies that we want to fight against. By believing that we must be subject to all of these difficulties, we limit ourselves to experiencing them.

We have absolute control over what we experience. It is all a reflection of our own consciousness. Nothing exists outside of our own Being. Once we recognize who we are in our true Being, we can command every situation in our experience.

Quantum physicists have proved that everything in our experience is a complex field of electromagnetic wave patterns and fluctuations that respond to our awareness. These wave patterns are conscious and have their own life force. When we give them our attention, they respond immediately by becoming situations that we perceive, feel and experience. The quantum field contains every possible experience in patterns of energy waves that we attract by our vibratory frequency, our personal energy signature. This is the energy that we align with in our thoughts, emotions and beliefs.

With our thoughts and imagination we create the forms of our experiences. With our emotions we create the content that we feel. It all arises for us out of the quantum field instantaneously in every moment. If we react to pain and suffering with feelings of pain and suffering, we continue to create situations that elicit those emotions. If we want to change our experiences, we have the free will to focus on higher-vibratory emotions, such as compassion and kindness. This kind of energetic radiance results in a higher-quality experience arising for us.

We can raise the quality of our human experience by raising the focus of our thoughts and emotions in our interactions with one another. We must choose the focus of our attention. By raising the frequency of our personal energy signature, we open ourselves to inspiring, loving and joyful interactions with one another, and we attract others living in the same spectrum of energy.

4.

Expanding Consciousness

Personal Expansion

Each of us has a unique presence of Being that is unlimited in every way. We are constantly created out of the quantum field of conscious plasma energy filled with the life force of universal consciousness and unconditional love. We are eternal manifestations arising out of the consciousness of our Creator. Because this is our reality, we can know all of this intuitively, and ultimately we open ourselves to being able to recognize our Being and realize who we really are, apart from our physical life on this planet.

We can come to this realization through inner silence and intending to be aware of the radiance of our heart energy in complete acceptance. In a meditative state we can focus on knowing intuitively while urging our emotions into higher and higher vibrations until we are filled with joy and love. The more we can be in this spectrum of energy, the closer we come to Self-realization.

Anyone can do this now. It is not important what our status in life is, or how we believe ourselves to be. We can ask within ourselves for whatever help we need and have it immediately, if we open our awareness to the energies of our guides and angels in a higher dimension of frequency. We have been held back from realizing our true Being by limitations that we designed so that we can be able to experience fully the tragi-comedy Earth life. We can now acknowledge those limitations and thank them for their service for the deepening of our consciousness. We can forgive ourselves for believing that we were mortal and could be intimidated by threats, disease, enslavement and torture. That level of vibration is so far out of resonance with our true Being, that it must be transformed or dissolved. By withdrawing our awareness and life force from it, we dissolve it from our experience. We could also transform low-frequency vibrations by modulating them into resonance with our intention, elevated emotions and imagination.

If we intend to come to Self-realization, we must be consciously open to the higher guidance that we all have always. We can recognize it intuitively by how it feels. We are guided in whatever direction we choose to go and however we want to do it. Our journey depends primarily on how deeply we align ourselves with the high-frequency energy of our heart and ultimately open ourselves to universal consciousness.

Living with Expanding Awareness

Although we can have intellectual knowledge of the magnitude of our consciousness, until we transcend our limited abilities in our personal experience, it's not really believable for us. We have been taught to believe that we are separate creatures with our own limited consciousness. Our limiting beliefs are so deeply held that we cannot even imagine that they could be false. They provide a comfortable perspective on life that we have grown accustomed to living with, even though they keep us locked into a fear-based understanding of life, filled with resentment, blame and a receding sense of freedom. The greater our awareness of the evil being perpetrated against us by our governments, the more angry, helpless and victimized we feel. Is this our true destiny? Can we transform our situation?

We are here on a journey to recognize our inner light. This is a journey that requires a conscious choice. It is a leap into the unknown. It's only partially unknown, because the quantum sciences have taught us that we all participate in universal consciousness that is the basis of everything, and that we are the creators of everything that we experience through our recognition of the energy patterns in the plasma field that envelopes us. Although the empirical world appears solid, its reality is electromagnetic waves that our consciousness interprets as our reality. Our essence is pure personalized conscious presence of Being. By holding visions and scenarios in our awareness we modulate the energy patterns in the quantum field, which provide the quality of our experiences.

By focusing on the experiences we are fearful of and do not want, we create those experiences in our lives. By focusing on what we do want with gratitude, we create those experiences for ourselves. We are all in this human experience together, and we share our creative energy with each other through our participation in universal consciousness. We get to choose the quality of energy that expresses in our personal lives, and we radiate this energy through our personal energy signatures with the power of our emotional focus.

A Perspective of Expansion

A perspective is a vibratory level that we choose to be present on.

Each perspective has its limits formed by our beliefs about our reality. These are our self-imposed boundaries of our consciousness. As we become aware of higher vibrations, our perspective changes. This changes our reality, because we're opening up to greater realizations, expanding our consciousness. We see all of the difficulties and suffering that people endure in the dimensional spectrum of energy that we have lived within. We've been here to experience the low vibrations of the dark side of life, so that we can understand the broad spectrum of energy that we did not know. We feel compassion for all of it, because we realize that we are all the same Being. Each of us arises from the universal consciousness of the Creator in the quantum field of being.

Each of us is a personalized presence of self-awareness with a unique energy signature, into which we project our reality with our thoughts and feelings. This is how we use our consciousness to bring the energies of the quantum field around us into alignment with our own, creating experiences that we live in and interpret through our own perspective.

We can elevate our lives by elevating our perspective. Instead of being against anything and giving it our life force, we can support the energy patterns that we want and give them our attention. We can become aware of higher vibrations by calling them from within. By intending to feel loving and compassionate in the drama of humanity's world, we are calling for higher vibrational experiences in our creative Being. This elevates the quality of our lives, which also makes it easier to continue to expand our awareness into higher-vibratory experiences.

As we are able to be more loving and joyful in our lives, we raise the frequency of our energy signature. Its radiance enhances the energy of all we encounter. In this way, we expand the consciousness of humanity as well as our own. We're opening our awareness to recognize our sovereignty, our essential Being, our infinitely powerful creative Selves, living in unconditional love, joy, compassion and abundance. We can expand our awareness into universal consciousness.

Creating a High-Frequency World

Is anything better than being filled with joy? Imagine being in the presence of our soul mates and twin flames. These are the ones with whom we share the deepest love and greatest attraction. When we're together, our presence is filled with joy and gratitude and love beyond measure. Imagine being with our soul family in deep, compassionate knowing of one another. Imagine doing wonderful things together and among groups of all these beings who love each other deeply. This is the world that is open to us through our recognition of it with our entire being. We can feel all of this within ourselves, and we can bring it into our experience by believing its reality and recognizing it as such. We are the modulators of energies, and we can radiate our visions and feelings into the quantum field for manifestation in our experience.

Manifesting our visions and imaginings happens naturally when we're in a high-frequency vibration of life. We can't force it with our will, our recital of invocations and prayers or any energy that is less than the high-frequency vibrations of our Creator through our heart. The benefit of doing all these things can be to direct our attention to transforming ourselves from being stuck in low vibrations of various shades of fear. We need to confront this energy from a higher perspective of compassionate wisdom.

We've limited ourselves to the frequency patterns of our ego. Without the guidance of heart-felt intuition, the ego has lived in fear of termination, and its constant quest is for greater fulfillment. By loving our ego and thanking him/her for taking us through our adventures into the low -vibration experiences of life and surviving, we can let those experiences be archived as history and forever remove our attention from them, giving them no more life force.

By resolving all of the limitations in our consciousness, we open ourselves to true freedom to know whatever we want to know in any moment, to feel however we want to feel, to be whoever we want to be. We open ourselves to knowing our deepest Being. This is our essence arising from the quantum field and offering us universal consciousness, which we've been shutting off from ourselves. Now we can transform this boundary and open ourselves to the unconditional love flowing to us constantly in our consciousness and vitality.

How Are We Multidimensional?

What is a dimension? It is a range of frequencies of electromagnetic wave patterns that constitute our experiences. It has upper and lower frequency boundaries. The dimension that humanity has been living in is the empirical world that we know through our senses and that we've been able to probe with our technology. All of these vibrations stimulate us in our emotional being in ways that we recognize through our feelings. Every frequency stimulates a unique emotion in us, which we can know when we are aware of this.

We can experience a range from very low-frequencies of deepest fear and torture to very high frequencies of ecstatic rapture. The times of shared beauty, joy and true love are rare in human experience. Most of the time we're around the median vibrations of our experiences. This is the primary resonant frequency of humanity's energy signature and is the focus of our dimension.

Life in a dimension lower than deep fear may be too horrible to endure. In frequencies at the resonant level of greatest joy and deepest compassion, we experience a higher dimension. This spectrum of emotion is not normally available for us and requires a leap in consciousness to experience. With a strong intention, any of us can do this.

We have learned that we prefer the higher frequencies of our emotions, the feelings of joy and love. As we intentionally align our awareness with these emotions, our visions also come into alignment with higher frequencies of beauty and brilliance. These are the qualities of energy available to us in a state of expanded consciousness in the next dimensions of higher vibrations. They come with extra-sensory realizations and greater power of manifestation into experiences. We begin to realize our life stream connecting us to the quantum field.

As we eliminate our accustomed blocks to awareness and other anomalous limitations, we're raising our consciousness into the next dimension, and we experience it more and more. Many of us are currently living in at least two dimensions and are intending to expand higher.

Our Symbolic Experiences

Our lives in this dimension seem real, because we recognize ourselves as we believe ourselves to be. In terms of physics, we are a presence of consciousness manifesting itself symbolically in a hypnotic dream. This dream is filled with metaphors and analogies of experiences. It is a never-ending and always changing tarot deck of symbology projected by our consciousness and appearing as our life experiences. Everything means more than at first appearances.

Symbols and music vibrate at frequencies that stimulate our emotions in ways that we understand. They stimulate us to create scenarios that affect our energy signature and, hence, the quality of our experiences. Their vibrations attract us to recognize their meaning and direction for us, if we are paying attention to them. If we look for the highest frequencies that we can imagine and feel, we will find our divine guidance through our connection with essential Being and our full consciousness.

Here we understand everything symbolically and communicate telepathically.

Every experience we participate in is a metaphor. Each experience has an energy signature that vibrates within a specific spectrum of frequencies. To understand the metaphors in our lives, we can feel the quality of energy present. Whatever spectrum of vibrations we're living in, the energy vibrations of the symbols resonate with this spectrum of energy. Our understanding must align with the energy of the experience. The metaphor makes this possible. It is from our higher guidance, if that's what we're looking for.

There is one all-important message inherent in all of our higher guidance. It is to pay attention to high-frequency vibrations of emotions and imagination within. Be in our expanded presence of Being in deepest love and joy.

Cosmic Consciousness

For those of us who have managed to escape from the hypnotic trance of humanity, even momentarily, we recognize that there are no limits to our consciousness. Whatever we wonder about or focus upon and feel, fills our being instantly. We can know anything and everything, because we participate in the conscious life force of the unified quantum field. We can just be a presence of self-conscious awareness, or we can express ourselves in any form, anywhere. This is a timeless experience. We can be aware of lifetimes on earth in detail and return to humanity's trance in a second.

Once we express ourselves within the limits of humanity's spectrum of energy, we do not allow ourselves to recognize or understand our true Being. We commit ourselves to the Earth experience with humanity in order to deepen our understanding of being. If we allowed ourselves to know our true Being, we could not recognize the low-vibration situations as real, and we would miss the whole point of being here. We are in a school of anomalous energy transformation.

There comes a time, though, when the game of limitation is finished. We're about there now. Everyone who's ready is now awakening from the trance. This is when we all expand our conscious awareness to our divine nature, intimately connected with the essence of our Creator. Once we realize the true essence of our Being, we can align ourselves with its high-vibration frequency. We are free spirits, sovereign and eternal, prompted by our intuitive knowing and the unconditional love flowing through our Being with our conscious life force. We feel each other's presence and quality of being, and we attract those who resonate best with us, and with whom we most enjoy being.

Elevate the Consciousness of Humanity

All of us who have achieved freedom from the hypnotic trance of humanity or can imagine doing so, can realize the essence of our Being by recognizing the vibrations of joy, peace, beauty and deepest love within our own consciousness. We can ask our divine presence to fill our awareness with these high-vibration emotions as we align our consciousness with the energy of unconditional love as much as we can. We can imagine that we recognize this energy enveloping all of humanity with the love of our Creator. This has a powerful effect upon the energy signature of humanity, affecting everyone, because it resonates with the energy of the unified quantum field.

The resonant frequency of the Earth is rising faster as we transform into Beings of divine love and joy. We're being enveloped by gamma ray clouds of photons and often blasted by them, as we can see on the Shumann Resonance graph. It is now more important than ever to Be in joy and compassion as much as possible and frequently attune to the energy of the Earth by being in nature, preferably walking barefoot, which connects us to the heartbeat of our planet and the energy of Gaia's heart. The energy signature of the Earth is rising faster than humanity's, and all of us will awaken from humanity's trance, except those who refuse. They will continue in this dimension, while everyone else moves into a wonderfully-refined life of truth and goodness.

Any of us can align with our naturally-high energy, which is our divine Self. This is the spectrum of frequencies that we enjoy the most. It is the natural energy of our own heart center. It is where we feel our life force flowing into us. It's where we feel enchanted within ourselves and recognize our expanded consciousness. We intentionally align all of our vibrations in resonance with our true essence that we know as our higher Self, our true conscious presence of Being. Through our higher Self flows the unconditional love that permeates the plasma of the unified quantum field, the expression of the consciousness of the Creator.

Assimilating Our Conscious Knowing

We may already have awareness of our intuitive knowing, and we've learned that we're living in a conscious simulation that we're creating constantly by our recognition of it. We're learning to align our perspective with the high-frequency energy of our heart, which has its own consciousness and lives to give us the life force that flows through us. This force has the feeling of unconditional love and connection with the quantum field and the consciousness of the Creator.

We've created many deeply held limitations of belief about our personal being. We're resolving these by recognizing them as they arise and comparing them to what we imagine our true Self to be. If there is any fear present, we present it with the love and compassion that naturally flows through us, and the confidence of our eternal Being. We may not yet have experienced all of this, but we will, if we intend to, because this is the energy of our true Being. We are not limited.

Our ability to realign the energies that we encounter enables us to transform everything into resonance with joy and love. We can do this only when we absolutely believe and know that this is how we participate in the energy patterns of everything that exists. This is our purpose. We are the creators of worlds, and together we are creating a new one for humanity.

We may not yet be fully consciously participating in the ascending and transforming energies of the Earth, but if we want to, we attract the energy that creates these experiences. We just need to recognize their reality. This takes practice. Once we recognize our inner guidance, we are on our journey to Self-realization. We need to be imaginatively and emotionally present and aware. As we use our intuitive guidance more and more, we gain confidence that we're getting the guidance we need in every moment. This is how we assimilate what we have learned, but have not yet become thoroughly enveloped in. With the will to process all of this, we are expanding and rising in frequency and raising the energy signature of humanity toward mass awakening to our pretentions and our real Being.

Penetrating the Matrix

We live in a unified field of energy patterns vibrating and spinning in alignment with universal consciousness. Our presence of being expresses itself as our personal energy signature in the field of all potentialities. This is also described as the vacuum fluctuation, where everything that we're aware of is produced by the spin of the energy patterns that we observe and interpret in our conscious awareness as material. Everything is a fractal of the consciousness of the Creator, right down to the sub-Planck particles and beyond, probably to infinitismality. It is the same in the other direction to infinity. We live within a very limited spectrum of energy patterns that we recognize as the empirical world. This is where the thoughts and emotions of humanity are focused. It is the backdrop to our conscious awareness and provides the limits that we operate within.

We have agreed to live within the human spectrum of energy and have accepted self-imposed limits to our awareness in order to make our experiences real for ourselves. We've gone so far as to accept the imposition of poverty, enslavement and mortality as qualities of our essential being. This is the Matrix that we live within in a hypnotic trance-like consciousness.

Quantum physicists have penetrated this spectrum of energy with experiments and mathematical formulae to find out what exactly it consists of. They have observed the consciousness of sub-atomic particles and many of the constituents of the cosmos. They have calculated the forces that hold the nuclei of atoms together and that keep our feet on the Earth. They have calculated the mass of everything from the smallest, faintest particles to the largest galaxies and have accurately solved the unified field theory. They have shown that universal consciousness is the basis of everything, and that we are participants in this unlimited consciousness.

The issue remaining for us is to realize our own unlimited personal consciousness within and beyond the Matrix. How we can do this is the continuing subject of these writings.

Directing Our Personal Evolution

Humanity is moving into two different paths in the evolution of our consciousness. One is an expansion of our natural awareness and the other is an adaptation to technological enhancement. The technological path involves artificial intelligence to largely replace aspects of our subconscious. It can greatly enhance our physical and mental abilities, but it is limited to the energetic boundaries of the empirical world. It has no emotion and has diminishing life force, because technology is an extension of our own being. All of the life force for a cyborg being must come from the diminishing presence of the naturally created human. The cyborg has no sense of self-realization and is not aware of the origin of its life force or having a conscious connection with higher-frequency vibrations. As the natural human fades into non-being in this path, the awareness of the emotional quality of life disappears. It is precisely the emotional quality that guides the way in the expansion of our natural awareness.

The expansion of our consciousness into infinite awareness is felt in our emotional being as we move into higher and higher vibrations of the energy patterns that we can focus upon. Our imagination follows with creative ideas for living. We can project our desire for higher-vibration energy patterns, which we will attract from a perspective of joy and gratitude for our present being. This perspective is a high-frequency energy pattern, which we radiate into the quantum field, attracting resonant energy patterns that become our experiences. This is an expression of our natural creative ability to modulate the energy waves in our present awareness, creating the quality of our experiences.

As we clear our emotional being of our limiting beliefs about our identity, we can practice and learn to be just present awareness, unconstrained by the body and human identity. We can remove our focus from everything and just be aware, preferably while grounded on the Earth in nature. We can begin to feel the energies around us and align our emotional awareness. All the energies and feelings are within our own consciousness.

By focusing on something from a negative or positive perspective, we give it our life force and enhance our experience of that quality of energy. Every challenge has a lesson for us. When we meet each challenge from a perspective of compassionate understanding, we bring the energy of the challenge into alignment with us, changing the experience. We are modulating the energy. This is our natural creative ability.

Beyond the Limits of Our Awareness

We've invented technology that has enabled us to perceive a much greater spectrum of our reality than our limited human senses have allowed. This has been possible, because inventors have had visions of possibilities beyond our normal experience. The same thing has happened in spirituality in terms of expanding our awareness. Although technology is not needed for conscious expansion, some can be helpful, such as biofeedback machines. By learning to control our minds and bio-rhythms, we can expand our awareness beyond the body. Many spiritual masters have taught exercises that enable greater consciousness. The result of having access to these developments is that we know much more about the nature of our would and our consciousness than our normal perceptions would allow.

Although technology is useful only in the empirical spectrum of electromagnetic energy patterns, spiritual awareness is unlimited in all frequencies. Technology has given us the knowledge of quantum physics, which has taken scientists to the limits of the mathematical and empirical realms. Beyond these boundaries is consciousness, which is the basis of everything.

Consciousness is who we are. We are pure awareness with unlimited creative abilities. Our consciousness is the consciousness of the Creator of all. We are creators of experiences. Currently we're performing in the drama of Earth human life, and we're taking this very seriously. Most of us are totally absorbed in it, so absorbed are we that we've forgotten why we're doing this, and that its all a play in our own consciousness. Everything that we confront and experience here is an extension of our own consciousness, and we control the quality of our experiences in every moment. We are the quality-control experts in our lives, especially when we realize how to do this.

We can realize that we are designed to be creators, just like our Creator. It's what we do, whether we realize it or not. We modulate the energy that flows through our awareness with our thoughts, emotions and intentions. We operate within whatever limits we have imposed upon ourselves through our beliefs about ourselves and our creations. The clearer we become in eliminating all of our beliefs, the closer we get to knowing our essential Being. It is from this perspective that we can express our Being with clear intent in the quality of our experiences.

Practicing Conscious Expansion

How much do we trust ourselves to be completely present in awareness? Only in this state of being can we sense our intuitive guidance in each moment. We can practice this. Ultimately it requires that we resolve all interference in our attention. We become mentally and emotionally neutral. We become open to what we intuitively know by the energy flowing through our Being.

Every vibrational pattern that we can be aware of stimulates the emotion that resonates with it. We live in an ocean of conscious plasma teeming with energetic patterns. We have the freedom to focus on any of them. Some of us can visualize images and forms, some hear music, and all feel the vibrations of everything we focus on. By expressing higher frequencies emotionally and imaginatively, we raise the vibrations of our personal energy signatures and attract higher-vibrational experiences.

If we encounter low-frequency energy patterns that feel uncomfortable and fearful, we can be aware with compassion and see the light within the situation, guiding it into resolution. We can be sovereign and free in every moment, always knowing everything we need to know for the enhancement of our present awareness.

Eventually we align ourselves so well in resonance with our intuitive knowing and the resonance of the Earth, that we are in the entire flow of the conscious life force vibratory spectrum.

We are connected in our essence with everyone and everything everywhere. Here is universal consciousness, unconditional love and eternal vitality. We can encompass all energy patterns in all dimensions in our higher-dimensional awareness. With this awareness we have the ability to modulate the energies that we focus on into high-vibrational situations. We radiate this energy through our energy signatures into the quantum field, where it attracts resonant energy patterns for us to experience.

Growing into Divinity

We can expand our conscious awareness as greatly as our beliefs will allow. Consciousness expansion begins in the heart. This is the portal through which our life force constantly flows to enliven us as self-conscious beings. There is a natural frequency spectrum of the incoming energetic patterns of being. We have the choice of whether we want to pay attention to these vibrations or disregard them and go off on our own adventure into a challenging realm. Most of us have chosen the adventure, but then we didn't remember what our natural attunement felt like or even believed we could be divine Beings.

If we can imagine being a fractal of our Creator Consciousness, we are expanding our conscious awareness to high-frequency wave patterns in the quantum field. Imagine Being aware of ourselves as grateful, compassionate, loving and joyful in all situations. We can look for the light everywhere and in everyone. We all emit photons that show our radiance. As we become more aligned with our natural energy spectrum, we begin to feel the deepest love enveloping us. We become aware of our true Being, arising out of the Consciousness of the Creator with all of our unlimited creative abilities, sharing the awareness of the light in all conscious beings with unconditional love.

Only by resolving all of our beliefs about our limitations can we begin to even imagine our true Being. Its state of Being is beyond the comprehension of our ego consciousness. We can begin to be aware of our ability to create emotions, and not just react with them to the energy confronting us in every moment. Our emotional creativity works in alignment with our imagination.

Our adventurous spirit can now hunt for our natural spectrum of energy, in which we feel joyful, expansive and inspired. We can search for higher vibrations in every aspect of life, which creates energetic patterns of high-frequency vibrations radiating into the quantum field for manifestation in our experiences. We become aware of wonderful things that are invisible to lower-vibration beings, as our conscious awareness expands, ultimately without limit. Our inner light grows brighter as we emit greater amounts of photons due to our receptiveness of unencumbered life force.

We Can Know Divine Guidance

We have the ability to transform all of our beliefs that are based in fear and that keep us from knowing our universal consciousness. We've adopted those beliefs about ourselves in order to participate in the vibratory spectrum of humanity. When we're ready to expand our consciousness, we can intentionally flow love and gratitude to our ego for giving us amazing experiences of low vibrations that we could never have had without this limited consciousness. We can love our ego beyond fear and into love and peacefulness. We can be aware of our natural intuitive guidance from the conscious life force that flows through our heart. This is subtle, high-vibration energy. In order to be aware of it, we must search ourselves for it. It is what we deeply know, apart from any beliefs about anything. It is a perspective of knowing in every moment. In this perspective, we can instruct and guide our ego in negotiating our interactions in the empirical world.

We can reclaim all of the life force that we've become accustomed to issuing to low-vibration agreements and fixations. In our natural state, we have an intense flow of life force, if we stay centered and grounded in alignment with the rising vibratory frequency of the Earth. We do not need to believe that we can get sick, age and die. These are all low-vibration energy patterns. These patterns are held in our awareness by our belief in mortality. Our beliefs are deeply held in our psyche, but we can recognize them and transform them into light through our gratitude for the experiences they allowed us to have, and that we no longer desire to have. Now we know what the realm of fear feels like and how it depletes us.

When we're ready to find fulfillment in everything, we can return to our essential presence of Being, beyond the empirical realm. This is where we can find our true passions and everything and everyone that we love the most. This is the natural spectrum of energy that we resonate with, when we realize what it is.

Identifying the One Conscious Being

In recognizing consciousness as the prime cause of everything, quantum physicists have realized that consciousness is mysterious and largely incomprehensible for us, yet we know what it is. We experience it as our awareness and our self-realization. It is our intuitive knowing and the source of our life force. It is our mental and emotional components and everything that we are that is not material or energetic. Our consciousness enables us to create forms and experiences out of electromagnetic energy patterns. It enables us to be creators of everything by our ability to modulate the ever-present infinite fluctuation of energies in the quantum field that envelops us. Our creative abilities lie in our mental and emotional expressions. We can pay attention to and envision whatever we choose, always. We can elicit feelings under all circumstances. Our life force radiates into the quantum field of all potentialities, from which it becomes the quality of experience that we focused our attention on.

Our consciousness is multi-dimensional, and we exist in other timelines at the same time as now, and are experiencing different situations and states of being. In our essence we are pure conscious, self-realizing presence of Being with infinite creative ability. In order to live within the limitations of the consciousness of third-dimensional life, we have compartmentalized our conscious awareness with limiting beliefs about who we are. We can examine each of these beliefs from the perspective of our intuitive knowing, conveyed by the life force flowing through our heart. It is a perspective of compassionate wisdom and knowing of truth.

We can become sensitive to our life force and feel its flow. We can observe how it changes in strength and feeling with the vibratory patterns that we focus upon. Low-frequency energy based in fear requires us to limit our reception of our natural life force and believe we are mortal. But our life force enlivens us eternally, and we enliven our body with the flow of our life force until we withdraw it. The quality of our body and our experiences depends upon the frequency of our energy signature, which we control through our thoughts, intentions and feelings. High-frequency energy, based in love and compassion, gives us joy, because it stimulates our intuition to know our eternal Being.

In our essence we are constantly arising out of the Being of the Prime Creator, the One Consciousness that is everything and everyone. It is beyond our current knowing, but it is us. We are fractals, personalized expressions of the Prime Creator. In our essential Being, each of us is Prime Creator.

And we believed we are just mortal humans.

How Can We Know the Unknown?

Perhaps we can start with what we actually know. If we depend only upon our senses, we know only the empirical world, which is where our traditional sciences are focused. If we add our emotions and mental processes, we enter the realm of our psyche. This is more of a semi-empirical, energetic world. If we move into our conscious awareness, we are in the realm of pure energetic expression. We all know that we are persons, but who is our person?

Every day we interact with the unknown, but we do not seriously consider it as such. We believe our world is solid, and we've learned how to make our way through life by moving from one established pattern of experience to another. Sometimes we'll venture out from our comfort zone with foreign travel or a new relationship, but it's all within the Matrix of human experience. The idea of having an experience that takes us out of our body consciousness and into the unknown realm of pure being is beyond our awareness. When it happens, however, we shift our consciousness to another dimension that is present everywhere, including here.

We have words and descriptions that allude to this realm, but it is beyond the boundaries of our current awareness and is thus unknown to us. Many who have returned from near-death experiences have described their experiences beyond the body. What is known is that their awareness didn't change when they were not in the body. They were still self-aware persons able to function fully in the realm where they were focused.

For us the greatest unknown is the depth of our own consciousness. We can create any circumstance that we imagine and feel. In our true Being, there is no unknown. We participate in universal consciousness and can realize our infinite creative power.

Just Being Ourselves

How high can we imagine ourselves going? How expansive can our conscious awareness be without fear? What can hold us back? We cannot make any progress in consciousness when we are desperate. We have created our own limitations to keep us from realizing how unlimited we are; otherwise, we couldn't be human.

The cosmos is ours to focus on, anywhere, anything and anyone. This is a process that happens in our consciousness, as we release all of our beliefs about being limited and mortal. Our bodies can be such, but our conscious personal Being is eternal and can express itself in worlds without end, and in more than one at the same time. The human can now be ready to transform. We are the Creator in our conscious life force, which flows to us constantly out of the Being of the Prime Creator. It flows throughout our personal energy field and expresses itself as us. We are persons, able to be anyone we want. We are Self-determined and have freedom of focus in every moment.

By learning to control our creative power from a position of zero-point awareness, we can be aware of our higher guidance through our intuition. We become just present in awareness, with nothing else going on in our minds or emotions. We begin to feel a desire for higher-vibration situations in life. Intuition is not intrusive, like the ego. To be sensitive to it, we must calm the ego with love and assurance of well-being. Then we can just be present and open our awareness through the energy spectrum of our heart—love, joy and ultimate service to our Source Being in gratitude.

We do not need any kind of personal assurance. Everything we desire comes to us. We do not need to hang on to anything, we just live by how we feel within and how we just innately know. We know that we prefer the realm of high vibrations. This is what we attract into our experience by following our higher guidance, free of ego and fear. We live in the dimension of peace, abundance, gratitude, forgiveness, compassion, love, beauty and joy. This becomes our real world. We can expand our consciousness as far as we want to go, because we are unlimited in our Being.

Penetrating the Depth of Our Consciousness

In our human experience we believe that we are separate beings with our separate consciousness. We have been taught this intuitively and verbally since infancy, and as a result we perceive that it is so. The empirical world is a realm of separate individuals existing in a spectrum of vibrations based on fear of termination of our being. It appears that materiality is the basis of everything, and that we are functions of our mortal bodies. But is this really true? And how can we know?

Quantum physics has shown that consciousness is the cause of all that exists. Countless experiments have proved this to be true. Everything from the minutest sub-atomic particles to everything that they constitute is conscious. There is only One consciousness. It is universal, existing everywhere in timelessness and expressing itself in a plasma field of limitless patterns of electromagnetic energy.

Our essential Being is constantly arising out of universal consciousness and expressing itself as our personal eternal presence of consciousness. We have the ability to materialize the energy patterns that we recognize, and we can attract energy patterns that vibrate in resonance with our own energy signature. What we believe that we recognize appears to us as material manifestation.

In recognizing ourselves, we are limited by our beliefs about who we are. Without limiting beliefs, we participate in the universal consciousness that is our essence. Because we have been taught exclusively all of our lives that we are limited beings, our beliefs are set deeply within the depths of our subconscious and require extraordinary experiences to clear them. The ultimate revelation is death of the body, but there are many lesser experiences that hint at realizing our universal consciousness. Some include near-death experiences, consciousness-expanding drugs, astral projection and soul travel, day-dreaming and night-time dreaming, and conscious regression experiences. We are complex Beings, expressing ourselves in different dimensions of existence simultaneously, with only a portion of our personal Being expressing itself in the empirical world. While we are focused in this realm, we are unaware of our larger Being. The key to realizing our true Being is transforming and resolving our limiting beliefs about ourselves.

Expanding Our Sense of Self

Most of us follow a routine that we've developed over many years. A few adventurers leave behind all sense of security and pursue their passion in whatever way moves them. Some dance, some paint, some go fishing, some play poker. Whatever it is, there's an inner urge that we all have that wants to blossom into full expression in a way that is unique to each of us. What holds most of back is a sense that we may fail in supporting ourselves and our families, if we do not work long hours at unfulfilling jobs that are essentially enslaving. The issue is one of self-identity, of who we believe we are.

Must we be subject to the dictates of political tyrants and the mandates of corrupt politicians or worse? Why are we subject to chaos, terrorism, child trafficking and other horrific evils? How do we extract ourselves and our families from poverty and starvation? Some situations may be so bad that we can't even leave our homes, or we wait at home to be terrorized, and some of us may not even have a home or any assets of any kind. We may be so sick that we cannot care for ourselves.

What is left for us to be grateful for, for this is the first step out of our predicament. We can release all resentment, hatred, depression, shame and attachment to anything, including our well-being, and forgive ourselves for our lack of trust in our true Being. We have the freedom of choosing to turn our attention within to our inner light, dim as it may be. There within our heart is our true radiance, shining with unconditional love and the power of universal consciousness, which enlivens all of us eternally.

Everything we perceive to be affecting our body is an energy pattern that we have attracted to ourselves by the focus of our imagination and feelings, prompted by our beliefs about ourselves. If we can manage to recognize that we have unlimited creative ability as the essence of who we are, and be grateful that we share in the essence of the Creator of everything, we can create a miraculous transition into a wonderful life for ourselves and our families, regardless of what is happening in the world around us. This requires intense and prolonged focus to overcome everything we've been taught to believe about ourselves, but it is the path of ultimate freedom and expanded consciousness.

Enhancing Our Inner Light

Humanity is in the process of awakening from the hypnotic trance that we have lived in for eons. We are being pressured more and more to accept total enslavement and impoverishment. The propaganda and mandates from our leaders are becoming more extreme, forcing us to realize that we are more than we believed ourselves to be.

We are following two different timelines now. Humanity's traditional matrix of control has become vibrationally unstable in the face of the rising vibrational resonance of the Earth. In their struggle to control humanity, they are becoming mentally and emotionally unstable, resulting in desperation. They need to keep humanity in a state of fear. This goes against the rising vibrational frequency of Gaia, Spirit of the Earth.

The natural frequency patterns of Gaia envelope humanity and require our alignment, if we want to keep living here. By aligning ourselves in resonance with the Earth, we are changing our perspective from fear to love and opening our awareness to a higher-frequency dimension of Being. Without the fear of humanity, the old order of control has no life force to provide amplitude for its vibrations, and it dissolves into unmodulated energy patterns.

In the interim we can choose to live in the love perspective, encountering all incoming energy patterns with love and compassion. This means that we are now being intentionally creative and not reactive in our encounters. This elevates the frequency of our interactions, resulting in our flowing into the higher dimension and disappearing from the low-frequency world. There is overlap, but everything proceeds according to our resonant frequency. We will have encounters that stimulate fear in us, as long as we recognize fear as real. It's not. This requires feelings in the spectrum of gratitude, joy, love, compassion and fun, as well as thoughts that are in alignment with this energy.

Regardless of what we encounter, we radiate conscious photons from our energy signature. This radiance attracts energy patterns in alignment with its resonance. By focusing our attention and emotions on high-vibrational situations and experiences, we can make the transition into a higher dimension of living, as soon as we give our attention to habitual high-vibrational feelings and thoughts.

Enlarging Our Field of Awareness

While we are embodied on this planet in an empirical world, our perceptions are limited to what we can sense. We interpret all of the energy patterns that we are aware of as solid reality. We've designed technology that expands our awareness into the subatomic realm and the ability to witness the presence of energy waves that we do not perceive physically. This extended our awareness into the realm of consciousness. We found out that the energy of subatomic wave/particles has its own conscious awareness beyond time and space. We found out that every change happens simultaneously everywhere, and we determined that there is a consciousness out of which everything is expressed. These are all observations of quantum physicists.

This universal consciousness envelops and pervades all conscious entities. This is our consciousness, when we can expand into it. Quantum physics has informed us that there is a universal consciousness that acts as a living Being, extending itself into all conscious entities that this Being creates, including us. Many people have called this Being, God or Prime Creator. Universal consciousness constantly creates energy patterns of every possible scenario in all dimensions, including ours. We recognize the energy patterns that we choose to focus on. As humans, we all agree to focus on the empirical spectrum of frequencies, which we've become so accustomed to, that we don't normally focus beyond them.

Shamans, spiritual adventurers and wild quantum scientists have stepped into a dimension beyond time and space. We know it exists, because of our technologically-assisted experiments. We accept these as empirical facts. This is where are current scientific knowledge ends. It tells us that there is a universal consciousness that pervades everything. It exists beyond time and space, and we can have access to it, because there is only One consciousness. It is our consciousness that we partitioned off in order to participate in the human experience. We can extend our awareness beyond our partition by remembering our true Being. This remembrance comes through the life force that flows constantly through the heart of our Being and gives us our vitality, awareness and essential Being. It is all beyond time and space, but within our consciousness. Our intuition can provide access to universal consciousness and higher guidance.

Choosing Our Quality of Life

In our personal consciousness we have chosen to be in our current situation in order to experience the energy that we are confronting. In the energetic dimension of humanity, it appears that we are living in a world that is outside of our own consciousness, because we have enclosed our consciousness in our physical bodies. Our bodies may be forcibly controlled by powers that seem to be outside of us, but our attention and perspective are uniquely our own always. There is no force outside of ourselves that can control our attention, which is our primary capability for personal growth and development.

We can open our awareness to a much greater truth about ourselves. If we are interested in expanding our consciousness, we can search our own sense of self for greater understanding. We can feel the quality of vibratory frequencies that fill our awareness. These stimulate our emotions and imagination. We can choose to focus on the feelings that we prefer.

We inhabit a quantum field of limitless energetic patterns and frequencies that we are not aware of until we focus on them. We have the free choice of which ones we recognize and allow into our realization, our emotions and imagination. These are the energies that become real for us in our experience. Other beings can project energetic patterns to us, but they cannot force us to focus on them. We recognize them only if we are open to them. We have absolute power over the quality of vibrations that we entertain in our awareness.

As fractals of the universal consciousness of the Prime Creator, we are designed to fill our awareness with the unconditional love and light that constantly flows through our essence, along with our personal vitality. When we choose to focus away from this energetic resonance, we feel unfulfilled in every way. We are well-acquainted with all of these feelings, and we know what we want. We just need to know that we have the innate power to create the life we deeply desire and to expand our consciousness beyond our human limitations just by being loving, compassionate and forgiving in all our encounters. By enveloping ourselves with high-quality vibrations, we transform our lives and the lives of everyone around us who is open to this energy.

Aligning with Higher Consciousness

In our true Being we do not need to do anything. We have the freedom to be anyone we want to be, including who we are as pure present personal awareness. We control the expressions of our consciousness through our imagination and emotions, just as we do now, but instantaneously, because in our true natural Being we are beyond the realm of time and space.

We live in, and we express ourselves, as a quantum field of plasma energy waves and patterns in all dimensions. We consciously inhabit one of these dimensions as our experience that we call reality. We've become enraptured within this dimension of every level and kind of fear, not knowing our true eternal Being.

In the empirical world we feel and imagine frequencies of waves and patterns of waves through the stimulation of our senses, which we project outward. We can use this ability in other dimensions as well, to modulate the wave patterns that we encounter into alignment with higher-frequency vibrations. We do this by how we feel and imagine ourselves to be in every moment. These are our limitations, until we disregard them by having a high-vibration perspective within universal consciousness. In this perspective we know intuitively everything that we need to know and act upon.

It is possible for us to have this perspective. It enhances our realization of Being unlimited in conscious awareness. In understanding the Matrix of human experience from a perspective of higher consciousness, we can be truly compassionate and kind to everyone we encounter who is unaware of expanding consciousness. All who are aware of our essence of Being can recognize each other upon meeting and know our gratitude and joy in our presence.

Realizing our unlimited nature happens to us when we align ourselves with the vibrations of the heart of our Being. This is who and what we are in the flow of conscious life force enlivening and vitalizing us in each moment. If we sensitize ourselves, we can feel and know this source of life within us. We can feel its vibrations stimulating love, joy and harmony in us. In every moment we intuitively know which vibrations we want to focus on, and we have the constant choice of doing so. Choosing to feel, imagine and act in alignment with high frequencies all the time is possible for all of us.

High-Vibration Inspiration

The quantum field consists of electromagnetic waves. These operate on intersecting planes of spinning wave patterns. Those that are magnetic are the ones we feel emotionally. Those that are electric stimulate our conscious mind, enabling it to recognize patterns and forms. We participate in interactions with both of these kinds of energy. Where the two planes intersect in our personhood determines the frequency of our energy signature. Our thoughts intersecting with our emotions creates a point in time that emits a radiation of electromagnetic wave patterns that we recognize as our world of experience. This process appears and disappears trillions of times a second. The process of our recognition and creation is happening in every moment, and it is always now, the present moment, continuously flowing into new experiences. Each now moment of our recognition emits a vibratory frequency that creates an experience that is energetically aligned. This changes with our mental and emotional vibratory focus.

The quality or vibratory resonance of our attention determines the momentary spectrum of our energy signature, which attracts compatible momentary energy patterns into our experience. There is a flow of frequencies, wave lengths and amplitude in every moment, guided by our attention and recognition. We feel these energies emotionally from deepest fear to greatest joy. Emotionally we always feel the energies that surround us, and we have the choice of aligning with them or changing our focus to align with a different vibratory pattern. We can be the masters of creation of our experiences, once we realize how it all works energetically.

By understanding energetic processes, our mind gains confidence that we actually can create miracles in our experiences through visionary focus on high-vibration scenarios. While holding such a focus, we can change the quality of our physical experiences, moment to moment.

In the intersecting point of mental and emotional planes of vibrations, every wave pattern stimulates an emotion in us. These feelings align with the visions in our mind. The vibratory spectrum that we feel and imagine in every moment radiates its energetic pattern into the quantum field and electromagnetically attracts energy patterns or experiences that align at the point of intersection of our thoughts and feelings, where they are synchronized vibrationally.

From this point of alignment of mental and emotional vibrations, whatever vibratory level we decide to focus upon determines the quality of our experience. By being clear and focusing into inspiration-expanding consciousness, we can open ourselves to recognition of a more wonderful and beautiful world of experience.

Aligning with Compassion and Love

As the resonant frequencies of the Earth continue to rise, it is important that we begin to realize that we all arise from and continue to be connected to the same universally conscious Being. In our current dimension of energy, we are embodied as individuals, and we cannot recognize our true Being, unless we are determined to do so. This experience of separation is intentional on our part, so that we can recognize and feel the energy patterns of separation from our Source. We have disabled our ability to know unconditional love and oneness of Being, yet we crave it in our deepest consciousness, because it's our natural state of Being.

Unity consciousness is the basis of compassion and true love. It is even beyond the feeling of closeness of spirit that we share with our dearest loved ones. We are all connected through the energy that flows through the heart of our Being. We are all the same One Being, expressing itself as all of our persons and every conscious being, every sub-atomic particle/wave, molecule, plant, animal, human, planet and galaxy. Even without personal awareness of this situation, we can know that we are all part of universal consciousness from the experiments of quantum physicists, who have identified universal consciousness as the source of everything.

Universal consciousness seems to come into expression and recede into unrecognized energy patterns many times each second. The entire cosmos is created and dissolved in infinity beyond time and space and comes into our awareness in time and space as we are able to recognize it. Each of us has recognition only within the boundaries of our beliefs about ourselves. These are the limits we have accepted for ourselves in our consciousness, in order to have a true human experience in the spectrum of fear. In our true nature, we could not know this experience, because we're in a dimension of unlimited energetic patterns arising in alignment with the love frequency from our own Being.

Limiting beliefs can be modulated into higher-frequency energy patterns that allow us to realize that in our essential Self, we are unlimited. We do this by confronting all qualities of energy patterns that we face in a perspective of gratitude, compassion and love. Even if we don't know what these emotions feel like, we can open ourselves to them, and they will come within us. Once we begin this process, we can just keep intending to be our higher Selves in every experience. This presence of Being expresses itself energetically in the flow of wave patterns at the zero point of the intersection of the planes of emotion and mind. This is our energy signature resonant pattern, by which we can modulate the energy patterns that we encounter into high-frequency vibrations in alignment with the vibrations of joy. We can always choose high-frequency energy patterns to recognize and experience.

When we can recognize that our conscious awareness is infinite beyond time and space, we can know that we exist eternally. We are the same consciousness as the Creator, with the same abilities of unlimited creation through our mental and emotional patterns and frequencies. Using our life force, we can love lower-vibratory energies into alignment with us, or dissolve their vibratory pattern. We can be open to even higher-vibratory energetic scenarios in alignment with our intuitive knowing.

The Potentials of Our Desires

Our desires are the perfect vehicles for creating our experiences. Desires are a melding of thought and emotion. When it arises from our heart, desire is an attraction to greater love and vitality in our life. Desire electrifies and magnetizes energy patterns that radiate into the quantum field that we inhabit all around

Chapter 4. Expanding Consciousness

and within us, and it interacts with the infinite variety of energies that we encounter in universal consciousness, creating previously unknown forms and levels of experience that are unique to us. In this way we expand universal consciousness. We are the creators of our experiences.

Because we eternally arise out of the awareness of the universally-conscious Creator, we are the modulators of energetic vibrations and forms through our own ability to focus on visions and feelings that we experience. According to quantum physics, the energy patterns that we recognize come into our experience. All energy patterns participate in universal consciousness and are aware of every conscious being that is aware of them. It is our awareness that changes the energy into material experience. If we are unaware of energy patterns, they are not part of our experience.

Because we participate in the spectrum of energetic patterns that humanity interacts with, we are always aware of all of these energies in some aspect of our consciousness, even if our ego consciousness is not, because we have compartmentalized ourselves with limiting beliefs about ourselves. As incarnated humans, we have endowed ourselves with doubts, fears and beliefs that create interference with the vibrations of the desires or our heart and keep them from manifesting for us, unless we resolve these through conscious intent. We can do this in many ways, some of which are deep meditation, hypnotic regression therapy, and through frequent and prolonged intense desire for being in the presence of ascended Beings who radiate great light and love.

As we resolve our self-imposed limitations, we gradually become aware that we are unlimited in our true Being. Our consciousness in the material spectrum of human experience is only an extension of our expanded consciousness. Our human consciousness is part of our creative essence and enables us to experience energetic frequencies that we could not in any other form. We are inhabiting a spectrum based on fear, so that we know that we will never again choose to experience this. We will have greater compassion for those who do, and we'll eternally realize the greatest love of the Creator, who has given us the free will to experience whatever we desire.

5.

Self-Realization

We Can Be Our True Selves

In our essence we are the embodiment of love in its most beautiful expression. Although most of us do not allow ourselves to realize this, the reality is present for us to know. This knowing comes when we align our sense of presence with our creative ability. What is real for us is what we recognize and believe in our conscious awareness. Our reality is created in our own consciousness.

We live in a spectrum of energy patterns that arise in the quantum field of all potentialities. This is a plasma of electromagnetic energy patterns formed in universal consciousness, which we participate in as personalized self-aware presence that expresses itself through the modulation of the patterns of energy in our recognition. The creative consciousness that underlies all of existence is our own consciousness. We are the creators of universes, and we have made aspects of ourselves small and limited in consciousness to experience life on this planet in a state of amnesia of our true being, so that we could experience the limitations in being that would be impossible for us in our natural state. From this state of conscious boundaries, we are now awakening and realizing our true Being.

Although we have devised a path of destiny for ourselves, we are free to change this at any time. We have gathered enough experience in our limited expressions of being to know the energies that we prefer to live in. We all want to live in the greatest fullness and expansiveness of life. We want lives filled with loving and devoted friends and the greatest love we can express and receive in our hearts. This is possible for all of us. It is the expression of our true Being. To experience this all we need to do is resolve and dismiss from our consciousness all of our fears and feelings of being less than who we really are. These are all anomalies in our energy signatures. If we can bring all of our energies into alignment with the highest and best feelings and thoughts that we can imagine, we can transform ourselves into our naturally unconditionally loving and joyous Being and live miraculous lives of fulfillment and beauty.

Self-Recognition and Realization

At some point we all realize that we are more than we have recognized ourselves to be. Often this doesn't happen until near death, when we become aware of our being beyond the physical body. Some of us naturally have abilities to travel in our awareness beyond the body. Some of us have the ability to read the thoughts of others, to feel the emotions of others, to communicate telepathically and to manipulate objects with thoughts. The great seers, magicians and masters have demonstrated unlimited creative abilities.

Why are the rest of us stuck in our physical presence and awareness? It is because we have chosen to experience the limitations of life in this dimension of energy, in order to experience what it is like to know and feel everything that is available to us here. This is where we can live in the low-vibrational world of the dark energy of fear, victimhood and limitations of all kinds. In our participation with the consciousness of humanity, we continuously create and recreate this world by our recognition, imagination and emotional involvement in it.

Our experiences here are all lessons in awareness and being. When we decide that we've had enough experience in this dimension, we're ready to awaken to our true essence. Knowing our real being would not allow us to experience the seeming reality of our Earth experiences. In order to perceive the empirical world as real, we have created limiting beliefs that enable its reality, and in this way our consciousness interprets this spectrum of energy as stimulation of our senses. We've done a masterful job at this.

Our consciousness is inherently unlimited, and is the essence of our infinite Being. Quantum physics has proved this on a micro level, and we can extend this to the macro level, because all that exists is energy, and all energy operates by the same consciousness. We express ourselves as configurations of electromagnetic energy in the quantum field of all potentialities. We can create whatever we focus our attention on. All energy patterns become material in our experience, when we recognize them. Every possible energy pattern exists in the quantum field, ready for our recognition and awareness.

By resolving all of our limiting beliefs about ourselves, and recognizing our participation in a higher-vibrational dimension of love, light and joy, we can realize who we really are and become masters of life in all dimensions.

Realizing Our Divinity

We all know that we are alive, at least in our imagination. If we have an out-of-body experience, we become aware of a larger sense of our conscious presence of being. We learn that we are more than our body consciousness. We all have the ability to imagine anything that we want, and we can also feel our emotions. These are all abilities that we can use intentionally beyond the body consciousness.

In searching for the origin of our consciousness, quantum physicists have determined that consciousness is unlimited and always present everywhere there is existence of any kind. There is immediate knowing of everything related to every entity in existence, with all of their thought forms and emotions. Consciousness is the essence of everything and expresses itself as the quantum field of all potentialities. Everything arises from the quantum field of universal consciousness and life force. This is the Source of our Being, and it expresses itself as the energy signature of each of us personally. Our personal presence of being, our self-awareness, is pure consciousness without embodiment. In our awareness, we can experience any being or thing that we can imagine. We feel the quality of every entity we choose to be aware of, and we are attracted to those of high-frequency emotions, the emotions of joy and gratitude.

Humanity has lived in a hypnotic state of consciousness in order to partition off our awareness by limiting ourselves to the empirical world, where we could pretend that we're mortal and can be enslaved and threatened in fear. Now we all have the opportunity to snap out of it and awaken. The realm we've been living in is becoming unstable and is disintegrating. It's being bombarded with high-frequency gamma ray photons that are drawing the frequencies of humanity and the Earth into a higher dimension of vibrations. This is now the natural flow of conscious life force.

If we can intend to recognize the presence of the Creator in everyone and everything, knowing that we are in essence all enlivened by the universal conscious life force and share the same Being, we can recognize the spark of the Creator Being in everyone. This spark of light is the connection of unconditional love. Once we can experience unconditional love, we know who we are. We are radiant Beings of deepest love and joy. This is our natural state of being.

Insights into Reality

Humanity has been under the delusion that we are our bodies, and that we live in an objective world that exists apart from our conscious awareness. Our thoughts and feelings have operated entirely within our imagined empirical world. We intuitively imagine and believe what other humans imagine and believe. This creates agreement on the nature of what we recognize as reality.

In order for us to move into high-vibration lives, we must transform our recognition of reality. How can we expand our awareness beyond this spectrum of energy in search of reality? What would happen if we imagine ourselves without a body and without a world as we know it? This is the state of just being present awareness, and is where we find the truth of who we are. We are a conscious presence expressing itself through our unique energy signature, which everyone in our presence can feel. We live in a quantum field of all potentialities. We can rest in our awareness of being. In Buddhism this is the state of the empty void. It is nothingness, but our awareness. Here we have no desires, nor expectations of any eventualities. In this state of being, we can feel our life force and unconditional love flowing into our being. We are filled with vitality and unlimited awareness, and we feel a deep connection with everything in existence.

We can interact with any entity existing in the quantum field. Any entity that we recognize also recognizes us, and we both appear instantly in form. Quantum physics has shown this to be true. We create our bodies constantly by recognizing them as we imagine and believe them to be, and our bodies recognize the essence of our being. Consciousness is the essence of everything.

Our being is creative with thoughts and emotions, and we modulate the energies that we recognize. We attract wave patterns that resonate with ours, and we recognize, feel and experience them. This is how we create our lives. Our potential is so much greater than we have realized.

Personal Resonance

On the path to awakening to our true being, we may experience significant inner dissonance. We have accustomed vibrations that are out of sync with our inner knowing and heart-felt emotions. We can transform our lives by bringing our lower-vibrational emotions into resonance with the energy of our heart, our deepest love and greatest joy.

All of our fears can now be faced, accepted, forgiven (they were the best our ego could do), and released with a refocus to joy in being. Our new perspective of being in joy, regardless of what else we may be facing, magnetically attracts energy patterns with compatible frequencies. These will be high-frequency experiences of joy and love.

The challenge of scrutinizing our beliefs about ourselves can result in our true freedom. Most humans believe that we are victims and slaves who are forced to obey someone, are mortal, vulnerable to many threats and other fear-based beliefs about ourselves. These are all low-vibration energy patterns. They can be brought into resonance with higher-vibration heart-felt energies by resolving our fears from the understanding of eternal, sovereign, unconditionally-loving Being. This is our natural self-realization in the quantum field, our true energy signature. This is the process of recognizing our true Self and realizing who we actually are, apart from humanity's beliefs about our identity. In our true Being, we cannot be threatened. We are sovereign and free to be completely Self-motivated in an energy field of unlimited diversity.

Chapter 5. Self-Realization

Part of our consciousness has closed itself from Self-knowing. We are now being motivated to take down our boundaries. We've set these up so that we could explore the dark side, if we want and find out how that feels. We've been free to explore the entire realm of low-vibration variations of fear. if we knew who we are, we could not do that, so we had to fake it by compartmentalizing our consciousness. Now the exercise of living in this dimension can be completed by resolving our misunderstanding of who we are.

The most well-known example of demonstrating our true Being without false beliefs about himself, is Jesus of Nazareth, who told us that he is one of us, not one different from us. He was able to transform energy patterns through gratitude, belief and inner vision. "I and the Father are One," He also said. The Father expresses Himself as the quantum field, as does the Mother. Our personal conscious Self expresses Itself as an energy signature in the quantum field. We are also One.

There is only One consciousness. It is universal within everything that exists. Everything arises out of the quantum field of all potentialities and receives conscious life force with unconditional love. Everything that exists has consciousness of its form and energy signature. We are the ones who can change our energy signature intentionally and jump to a higher spectrum of frequency and being. We can examine and update our beliefs about ourselves. We can not only resolve and transform our beliefs, but also free ourselves from any conscious limitations, which we have imposed upon ourselves, in order to share the human life experience. Now we can move to a higher dimension of vibrations by raising our focus to high-vibration visions and emotions and intentionally staying in that frequency spectrum. This will bring our energy signatures into resonance with our true Being.

Resolving Low-Vibration Energy

We have deeply-held fears that we must resolve, because they keep us from feeling the energy of our heart, and they have made us feel unworthy of deep love and joy and prosperity. We can't drag our identification with these energies into a higher dimension, but we can be open to new feelings that are better than we're accustomed to.

If we identify ourselves as unworthy, our subconscious does not allow us to experience that higher frequency. Our frequency needs adjustment in our belief about ourselves. Our frequency is a combination of energy patterns of what we see, feel, think about, imagine and experience. We can choose what we focus on, think about and imagine. We can be aware of our feelings all the time. They recognize the quality of every experience and can be used to change our vibratory frequency. Resolving our unworthiness and changing our perspective to Joy is life-transforming, If we want to do it.

Feelings of unworthiness, shame, anger, and victimization all originate from the compartmentalization of our consciousness. We believe that we are the confined beings that live in our bodies and are subject to all manner of indignities and ultimately death. Our bodies are an expression of our consciousness and are continuously created in each moment. We needed these low-vibration beliefs to enable us to explore dark energy in its most convincing way. When we feel that we're ready for the truth of our Being, we can critically examine those beliefs and open ourselves to expanded awareness.

From the perspective of quantum physics, what is needed is focus on higher visions and emotions. We are completely self-determined beings. Our perspective and the quality of our experiences are completely created in our own consciousness. We can raise the frequency of our energy signatures. We can intend to be compassionate and kind in interacting with all beings. We can intend to recognize the divine spark in everything and everyone. We all arise from the quantum field and are personifications of the same Being, expressing Itself as a plasma field of unconditional love, abundance, vitality and universal consciousness. By loving our bodies and even being in awe of their beauty, we can bring this energy into our experience.

We can feel the presence of our complete Self through the energy of our heart and know its guidance through our intuition. This is who we are in our deepest Being.

The Eternal Now

The quantum field, which is the source of our being, exists beyond time and space. It is everywhere in every moment. Every moment is now. From our earliest training in this lifetime, we have learned to think about our life differently from this reality, because human consciousness is intentionally limited within an artificial construct that we have all agreed to participate in prior to incarnation. We take on conscious amnesia upon incarnating, so that we can get the full experience of this spectrum of vibrations. We soak up all of the social conditioning, even though it often doesn't feel good or make sense. We learn fear of all kinds. Without fear, we would naturally align with our higher resonance. At that level of awareness, we can begin to examine all of our assumptions and beliefs about ourselves.

In being aware of our feelings, we can imagine high-vibration situations that we would love to experience. We can feel that we are in those situations. By being mentally and emotionally in the high-frequency spectrum of inner or outer experience, we can make a jump in consciousness to a higher dimension of life. Even a glimpse of a society of loving, joyous, kind and compassionate beings is a strong attraction for more conscious, emotional contact.

Time exists only within the human ego's construct of reality.

We were designed to encompass universal consciousness with the ability to know everything everywhere that we give our attention to. We exist outside of time and space in our presence of being consciously alive. This is our true Self. Here our awareness opens up to a higher dimension of experience. Here we are sovereign in our own Being, eternally conscious and Self-aware. We are filled with unconditional love for everyone and everything in conscious connection to all beings through the quantum field of conscious vitality.

Who We Are and Why We Are Here

We are our conscious presence in a sea of creative, living plasma, containing every possible pattern of energy. We are undefined. Our awareness is infinite, if we do not limit ourselves. In our essence of being we have no form. We can express ourselves in any form, limited by our beliefs about ourselves.

In our current human awareness we have projected ourselves into our situation now. We have been given the knowledge and intuition to be able to handle all of our challenges. Our experiences include challenges that we probably would not have chosen to have to face, but they are intended to stretch our awareness of our abilities and gain a deeper understanding of ourselves.

We are becoming aware of the metaphorical nature of our focus as reality. We had to give ourselves convincing limitations in order to make our experiences real for us. We have intentionally created all of our limitations, especially our beliefs about ourselves. These are very deeply embedded in our being and require the light of our attention and the love in our hearts. We're taking our ego consciousness into a higher-vibration experience, in which we relax in absolute compassion and love. We allow ourselves to be content in our being of unconditional love and joy. This may be a leap, but we can do it with the will to be at peace. As our beliefs come into our awareness, we can introduce them to a new perspective. With practice we can stretch our awareness into resonance with the energy of unconditionally-loving life force flowing through our heart.

Each of us incarnated with a life intention of expanding our awareness of our Selves, and in this way also raising the consciousness of humanity. Once we have an encounter with our true Being, and we feel Its magically wonderful essence, we are expanding our awareness into a higher realm of joy and beauty.

Everyone who is drawn to this resonance of being is moving into a higher dimension of heart-felt living. We are being attracted to one another, and we recognize each other through the radiance of our eyes and the connections of our vibrations.

Creating Our New Lives

We all want personal fulfillment. We want wonderful relationships with friends who are compassionate and loving. We want abundance in all areas of our lives. We want vibrant health and beauty. We want to express our passions and dance with the wind in ecstatic union with great joy. These are now all possible for all of us.

Our creative essence is unlimited in every way, but we have designed and implemented powerful limitations for our Earth experience with the energy expressed by humanity. We have lived within these limitations for thousands of years, unable to realize our true being, but now we have the opportunity to transcend all of this and transform our lives with the creative life force that flows through our hearts from the quantum field of all potentialities. Our Creator is sending vast numbers of high-frequency gamma-ray photons through the central suns of our universe and galaxy to activate our potential for the expansion of our consciousness beyond our self-imposed limitations to realize our true Being.

We can align with the rising resonant frequency of the Earth (as shown in the Shumann Resonance graphs), we free ourselves to become more expressive of our own love and compassion. We begin to recognize our innate ability to raise our vibratory frequency by intentionally focusing on the great love and support that we feel through our heart.

As we expand our inner light, we move beyond the threats and troubles of the lower frequencies in human society. We can drop our entrainment of poverty and enslavement and free ourselves to live and express the desires of our heart and expand our awareness into a world of beauty and wonder. We can call on our angels and guides and our true inner Being to inspire us to express the greatness that is within all of us. We are creating a new reality of experience in a higher dimension of abundance, peace and joy beyond measure.

The Flow of Life

Universal consciousness pervades the entire cosmos and informs everything that exists from sub-atomic particles and waves to galaxies and universes and all entities everywhere. Consciousness is the basis of everything and everyone. This is the basic recognition in quantum physics. It is the expression of the Creator. Every being has unique abilities and attributes. All is designed to flow and interact effortlessly, manifesting beauty and elegance everywhere. Photons act as if they are a single entity. They know instantly always what every other photon is experiencing everywhere. Perhaps humanity has this ability within its collection of beings.

What happened to humanity? Why is life so difficult for us? Can we get realigned to live effortlessly? Because humanity chose to live within the limitations of a low-vibration spectrum of energy, we inevitably suffer in order to have a deeper experience of life's possibilities. When understood from a high-vibration perspective, we have gathered much wisdom and compassion, deeper than would have been possible from an expanded perspective.

Our realignment happens when we intentionally seek high-vibration energy in ourselves and in our relationships. We feel gratitude for our current situation. We are consciously alive and have the innate ability to create whatever we want from the energy of our heart. We must open our awareness to the vibrations of joy. We are creating our own path to awakening by recognizing the reality of a higher dimension, where we are unlimited. Confidence is needed here to the point of absolute belief in our own higher Being, whom we all innately know, if we seek this expedience.

Choosing to awaken to our true Infinite Self, requires a state of objectivity toward life. Our guidance comes in the moment when needed, without holding onto anything from the past. We all have our own inner coach. We are designed to naturally desire life in a higher dimension. In alignment with our Source energy, coming through our heart, we can make the leap in consciousness that allows us to break out of our trance in humanity's dimension. We intentionally align ourselves with the vibrations of nature, our alignment with Gaia. We can imagine and feel the vibrations of life force coming through us with the unconditional love and consciousness of our Creator. This is our true Being, our own higher Self.

Finding Our Inner Light

Our only reality is within. Deep in our sense of being we know our participation in the consciousness of the Creator. Within this consciousness we exist and express ourselves as patterns of energy comprising our personal energy signature that exists eternally in the quantum field. We can create anything and anyone that we express through the energy of our heart, with our imagination and emotions. We radiate our vibrations into the quantum field for manifestation in the dimension intended.

We created the empirical world to experience a physical presence in ways not possible for us otherwise. Unfortunately we were able to create the personal boundaries and limitations necessary to participate in this energy spectrum, and they were so convincing, that we got our attention locked into these limiting vibrations, unable to be aware of our true Being.

Our limitations are all self-imposed in our own personal consciousness. We have taken on these limitations intentionally or by our acceptance of them. Many have been accepted through trickery and deceit. We believe them all, and they require us to believe that we are inherently limited, mortal beings. The limitations are all based in fear. We believe that we can be threatened, intimidated, tortured and killed. We believe that we have flaws, are ashamed of ourselves in some ways, and must depend on others for our survival and well-being. These beliefs are all false, as we find out, if we can penetrate our consciousness deeply enough.

The first requirement is for us to be in joy. This is the level of vibration that takes us into the energy of our heart. Joy is accompanied by gratitude, love, peacefulness, and compassion. These are all natural feelings in our true Being. By asking for them to arise and imagining situations in which they would be present, we can raise the frequency of our energy signature. We can resolve all of our limitations and make the leap into a higher dimension, where we are present as sovereign, eternal, infinitely powerful creators within the universal consciousness of the One Being that we all are.

Excitement About Our Passions

As we raise the vibrations of our personal energy expression, our emotions become deeper and more meaningful. They participate in the rising frequency of our sense of being. Everything becomes clearer and more intense.

Any anomalies become glaringly obvious, asking for our love and compassion. With a high-vibration perspective, we can resolve and transform every situation that we encounter in our physical experience and the imaginings of the ego.

In our true Being, we can extend our essential presence infinitely without time or space. Our life within the spectrum of vibrations of humanity is just one of an infinite number of possible expressions of our Being. One of the benefits of being human is the development of our passionate emotions. We have the ability to know everything we need to know in any moment through our heart-felt emotions that feed our intuition. We can understand situations and circumstances with compassionate wisdom.

In the world that we are creating with out vibratory radiance in the quantum field, we are free to develop our high-vibrational passions without interference. We can practice this with abandon! We can recognize emotionally the divine light of life in every being we encounter. This is the energy we can interact with. Very low frequency beings will not want to be in our presence, unless they choose to realign their vibrations with ours. Low-vibration beings and things are becoming energetically unstable. The Matrix we inhabit, which is constantly created by the consciousness of humanity, is weakening, as humans begin to open up to a greater reality. We are being directed to align with our true Being in every circumstance. Once we drop all of our pretenses and false beliefs about ourselves, we can begin to intuitively know our true Being in the radiance of our heart energy.

Our Infinite Being

In our conscious awareness, most of us believe we are merely human beings. We deeply believe we are limited to our physical bodies. Some of us have escaped from our physical awareness into the astral realm in our conscious awareness. Our inner sensitivity can help to keep our vibrations high. This is the alignment in the vibrations of our energy signature with the heart-felt high-frequencies of our natural rhythm. We feel it deep within. It is our natural expression of our true Being in our feelings of joy and love and our freedom of eternal Being. As we are able to resonate in this spectrum of energy, we free ourselves from the limitations of human consciousness, and a great freedom of awareness opens to us.

We become aware of the subtle flow of life. We can direct our awareness anywhere and know everything we want to know instantly. This awareness is not dependent upon the physical brain or heart. This is the awareness of our Being, of which our human consciousness is an expression intended to enlarge our experience.

We are part of the quantum realm of energy, universal life force and consciousness. We are just Being present awareness. We are our Self-awareness, which includes every ability we can imagine and feel. We arise out of the consciousness that expresses itself as the quantum field. Our consciousness is unlimited in the expressions of our imagination and emotions. We can stretch our conscious sensitivities to the super-minute levels of Plank Particles, sub-Plank, sub-sub-Plank *ad infinitum*. These are trillions of times smaller than a proton, and are fractals of the proton and the entire cosmos, as are all other entities. All participate in the universal consciousness of the Creator of all. This is the Consciousness that empowers and enlivens the unified quantum field of all potentialities and provides the conscious life force of all beings. This is the Source of our Being, our divine Self. In our essence, this is our deepest and greatest awareness. It is timeless unconditional love and joy. It is awareness of infinite Self-realizing Being.

Recovering our Self-Identity

Each of us is an expression of the universal consciousness of the Creator as a unique personal creative genius. Since we have made our way through life on this planet under the identity of our limited ego consciousness, we have not known our true Being. We are present in our awareness to create experiences for ourselves and our Creator in alignment with our personal genius out of the infinite possibilities of existence. We are here now among humans on this Earth to awaken to our true nature far beyond the imagination of the human mind.

It can help us to understand what quantum physics has discovered about the nature of the universe and its basis in universal consciousness. These are no longer concepts that only spiritual adepts know. This knowledge is now scientific realization. As more physicists grasp what they have discovered, they are gradually realizing that we are living in an energy field held in universal consciousness by the collective consciousness of humanity. It has no substance in the way that we have learned to recognize it. For us the material world is only an interpretation of our conscious awareness of the particle/wave patterns that we all create together with our self-identification with them. Our beliefs about ourselves keep us confined within this spectrum of energy.

Without fear, we have no need for limiting beliefs about ourselves. Fear is a low-frequency vibration that permeates all energy fields having a frequency below that of unconditional love. The realm of fear encompasses all consciousness that does not recognize the Source of our Being. Limited self-consciousness feels that it is the source of its own life. It is mortal and can suffer. This has been the conscious state of humanity, limited by fear.

It usually takes a leap of faith to abandon the traditional human conceptual framework of reality to open our awareness to higher-dimensional being. We're working on being able to make the transformation in easier steps. By realizing how quantum energy works, we can see the changes that we need to make in our lives as patterns of energy fluctuations, which we can control and align with our emotions and imagination. Our process of transformation can be realized as energetic expressions that we live in through our perspective in thinking about things and our reactions to situations that we face.

All of our thoughts and feelings have energetic expressions in the quantum field. By raising our predominant vibrations of what goes on in our minds and feelings, we can raise our personal vibratory frequency, and we can do this as much as we're aware of it. In this way we can keep going higher in our life experiences while expanding our conscious awareness, until we become self-responsible and creative of everything we need and want. We come into alignment with our divine Source energy of unconditional love, joy, peace, abundance, freedom, sovereignty and eternal Being.

Transforming Our Limitations

The only power that constrains us within the vibratory spectrum of humanity is self-imposed. Our limitations are concealed deep in our subconscious. We designed the world we live in to be as convincingly real as possible, which means that we did not intend to be able to transform our situation, unless we are soul wanderers and have experienced life beyond the spectrum of humanity's manifestation. The only way to know what's beyond our human limitations is to recognize what is beyond our current recognition. This may prompt us to search our consciousness with emotional awareness.

We can imagine just being present in the unlimited quantum field of all potentialities. There is only universal consciousness, which we participate in. Everything we perceive is a reflection of our own consciousness, which we choose to pay attention to.

Our mental and emotional focus radiates the vibratory pattern of our state of being into the vacuum fluctuation of the quantum field, where it attracts resonant energy patterns that become our experiences.

As we become aware of our abilities to create our lives, we can take ourselves into extraordinary experiences. High-vibration energy always is of love and joy, compassion and gratitude. These are the energies of our essential being. While we are intentionally experiencing this perspective, we can confront our false beliefs with compassionate understanding and resolve them.

When we have resolved and relinquished all limitations, all false beliefs about ourselves, which have been necessary for the reality of our human experience, we become receptive to higher-vibrational insights and experiences, which come with the life force that flows through the energy of our heart. This is the source of our higher guidance, which we can rely on for the truth of everything, once we train ourselves to be alert to it.

Expanding Our Personal Abilities

By expanding our awareness beyond the limitations of our physical bodies, we can believe and deeply know that we are eternal and unlimited in our consciousness. For most of us this involves practicing conscious breathing, transcendental meditation, using biofeedback devices, yoga, Eckankar and other techniques that help us realize our capabilities. Our intuitive knowing becomes experiential knowing beyond our human limitations.

As our expanding awareness opens to recognition of our participation in universal consciousness, we realize that we can recognize the energy patterns of all that exists and the essential being of our Creator. Our belief in separation from one another and from the Source of our life dissolves in our alignment with divine love. We are becoming our true Selves, able to understand

and realize our creative ability as unlimited. We are the personalized essence of unconditional love and joyous being.

Conflict within ourselves and with others becomes impossible, because we experientially know the truth of our Being. We no longer have fear or unfulfilled needs. The grand experiment of living within the limitations of human consciousness is changing, as the Earth rises in resonant frequency, and blasts of conscious gamma ray photons from our galactic core upgrade our DNA and enhance our conscious abilities.

With our knowledge of quantum energetics and our ability to transcend our limited perceptions, we are transforming the energy spectrum of humanity into a higher dimension of living. Human life is to be no longer about survival and enslavement, but a celebration of unconditional love, joy and beauty in infinite creative potential and realization.

Opening to Our Inner Light

The light of our soul dwells in our heart. Our physical heart is symbolic of the essence of our Being, and it is the organ through which our bodies are enlivened. Our life force is constantly distributed by our heart throughout our body in combination with the conscious life force that flows to us through our breath.

In our human condition, we shut down our receptivity to much of this life force energy through negative thoughts and emotions and shallow breathing, as well as disconnecting ourselves from the natural vibrations of the Earth and our cosmic environment. Fear of threats in our social and natural environments have kept us in emotional stress. Our egos are hiding unresolved wounds, especially from childhood, when we are most vulnerable. We've developed many false beliefs about ourselves as a result of our experiences without higher guidance. All of these things and more make us spiritually dysfunctional and unable to enjoy the power and awareness of our true Being, but we do not have to be stuck in this condition.

The first steps toward recovery of awareness of our expanded Self and essential Being can be taken by focusing on our current emotional state and recognizing the quality of the energy that we are holding. We have free will to choose what we feel and think about. We can recognize when we're reacting irrationally due to hidden emotional trauma. At these times we can ask for guidance from within to identify the source of the trauma and resolve it with love and compassion for our inner child. As we clear our consciousness of our addictions to low-vibration energies of fear, greed and lust of all kinds, we can open ourselves to the flow of the natural energy of the Earth, beautiful music, deep inhalations of vibrant breath and the realization of our own conscious presence.

We have much to resolve within our psyche, but it is all self-imposed and can be self-resolved. Sometimes this requires extracting ourselves from everything we're involved in and starting over. It requires sensitivity to our intuitive knowing and the willingness to receive inspiration from our heart through our emotions and imagination. It is being open to our deepest knowing as we gaze intently into our eyes in a mirror to see the light and feel the love coming back to us. In our true Being we are complete in our Selves and radiant with the energy of love, compassion and joy.

Finding Our True Nature

Our self-identity has been shaped by complex energies in our continuing awareness. We have imagined and allowed discordant vibrations to disorient us. We have lost our alignment with our true being. In the process we've scattered our life force into the realm of mortality and separation from our self-knowing, leaving us with very little personal power and few abilities. By believing that we belong in a state of enslavement, because we create imperfection, we have trapped ourselves in a state of limited awareness. Once we are aware that we're involved in a limited spectrum of experiences, we can begin to realize that there is much more beyond.

Our intuition can align us with our natural vibrations. We may recognize the conscious creative energy that directs everything, as the supreme Creator of all, from whose Being we all are continuously created. This universal consciousness expresses itself as the unified quantum field of all potentialities, the vacuum fluctuation. Out of this realm of infinite patterns of energy, we express ourselves by constantly radiating the energy of our being into the quantum field, within which we reside as fractals of the One universal consciousness.

To recognize our most consciously expanded presence of Being, we can stop focusing on low-vibratory experiences and reclaim our life force, so that we can be empowered in the essence of our being. We become like energetic capacitors, directing our conscious focus on high-vibration experiences only, while being open to act intuitively in every encounter. Our true intuition is in the flow of conscious life force constantly running through our awareness. We only have to recognize it and realize what it is.

Through intentional alignment with the feelings within our life force, we can expand our awareness beyond the matrix of energies that we share with humanity. We can begin to observe what happens to us as we focus on different frequency spectra. We may find that our focus and emotional state create the quality of our experiences. As fractals of the Creator, we are unlimited in our ability to create everything. We modulate the energies in the quantum field with our energy signatures, which resonate with our thoughts and feelings. We are creating a new world with our vibratory state of awareness, a world that is natural to every enhancement of life, vitality and expansion of high-vibrational experiences of love, joy and abundance.

Conscious Mental and Emotional Awareness

What is consciousness, and how is it different from our mind? Consciousness is self-realization. The mind arises from consciousness and is the thinker of thoughts and concepts. Consciousness is the essence of our being. For each of us, it is self-awareness as a personalized eternal presence of being. It is the force of our life and the creative essence from which all attributes and abilities arise. It is the creative cause of everything. We are consciousness, sharing awareness with all that exists, to the extent that our beliefs allow us. We are fractals of the universal consciousness, being identical to the Creator in every way. In our complete Being, we are Creators, having infinite abilities in every way.

Our mind facilitates the expression of our beliefs, keeping us from realizing our complete essence. All of our limited beliefs about ourselves have been formed without our intuitive guidance, which comes from our consciousness. We have given control of our consciousness over to our mind, which controls our thoughts. Our mind does not know how our emotions are the empowering creative energy that provides our experiences. Our emotions arise directly from our consciousness, not from the mind, but they can both affect each other. The mind has blindfolded our consciousness. By removing the blinders, we achieve self-realization of our complete essence.

The virtual reality Matrix that we have lived in together is an illusion. The way of understanding how to participate successfully in this Matrix, is to feel the vibrations in our present moment and imagine the quality of energy. By recognizing the vibrational level of the energy in our presence, we can choose to change it. We have that ability. We are modulators of energy patterns. We can change the vibrations of our emotions and our thoughts, which affects our personal energy signature. Our personal energy signature radiates conscious photons into the quantum field that envelops us, creating the quality of our experiences. We are beings of light, beyond the material virtual-reality world.

In our current incarnation, our personal truth is present in our awareness coming through our heart as our intuition. This is where our mental attention can be peaceful and gracious. In opening our awareness to our intuitive knowing, we are aligning with the vibrations of our essential Being. We become expressions of the divine Creator.

More Intimacy with Our True Self

We may want to know who we truly are, but we don't know how to go about it. We've designed the Matrix of human experience to be so powerfully engaging in our consciousness, that we may not even suspect that we are anyone more than our empirical life. If we look beyond the obvious to the higher-frequency electromagnetic wave patterns, we can feel them.

In using the intentional power of our focus, we can be acutely aware of our feelings in every moment. We can learn to be aware of the vibrational patterns in our experience through how we feel. We can feel the difference in the energies of fear and those of love and compassion. When we are aware of the emotional level of the vibrations that we encounter, we can attract and be magnetically attracted to the spectrum of energy patterns that resonate with us. By being who we truly want to be, we learn to direct our focus to the energy patterns that we love the most.

We may encounter many self-imposed blocks to higher awareness that we must resolve in order to be free. Our ego cannot be eternal if we believe that we're mortal. The ego was created to believe that it's mortal. It's the only way that we could be fully human. The ego, however, must respect higher guidance, which we know intuitively in our Being. Our ego needs us to love him/her and be considerate and understanding. Only when our ego is without stress can we align with the intuition of our essential Being. The ego can learn to trust that we are aligned with our true essence that constantly flows to us through the energy of our heart in our intuition. This is where we know ourselves beyond belief and beyond empirical experience. We are each a pure presence of self-awareness within universal consciousness, having infinite creative abilities, one of which is the expression of our energy signature as our physical body. Without beliefs in our limitations, our bodies can be eternal, youthful and beautiful, if that is how we would like to express ourselves. We do not need our bodies, but we are now capable of truly enjoying them.

We can open our awareness to this higher dimension of vibratory frequencies, which we can immediately recognize through our feelings. When we choose to focus in the spectrum of love and compassion, we are moving into an intimate alignment with our true personalized essence of Being.

Living in the Energy of Self-Realization

If we have resolved the major blocks in our consciousness and are able to be still and aware of our inner promptings, we become able to create our new reality in the higher dimension of frequencies. We are able to keep our perspective in the spectrum of love and compassion as we face every situation that confronts us. We know that every obstacle in our lives is placed there with the conscious intention to awaken us to our reality as multi-dimensional Beings. If we don't know this lesson, life becomes still more difficult, until we reach desperation. Then we either make a leap in conscious and align our thoughts and feelings with high-vibrational awareness or terminate.

This is what's happening as the resonant frequencies of the Earth keep rising. Any discordant human energies become unstable. Their electromagnetic wave patterns receive constant interference from the Earth. We must align with the rising frequencies or get ready to leave.

If we can stay in high-frequency thoughts and feelings, our lives become wonderful, because we magnetically attract resonant frequency situations and people. We repel all low-vibration energy, and it does not enter our experience. We live in a world of personal experience of abundance and well-being. Our friends are kind and thoughtful.

We are naturally creative. Regardless of our state of consciousness, we are still creators of the quality of our experience by the spectrum of vibrations of our energy signature. We have created a spectrum of energies within which we vibrate. We keep our thoughts and emotions within the boundaries that have been accepted by humanity. The only ones who venture outside of those boundaries are adventurers, seers, mystics, inventors, visionaries and the insane. Those who seek Self-realization are outside the boundaries. We are the cosmic travelers, seeking to be more transparently light and love wherever this level of vibration attracts us. There are many of us now, and more are awakening. We're creating and moving into a higher vibration of experience and a greater realization of our true, infinite Being.

Awakening from the Human Virtual-Reality World

Life holds many complexities to keep our conscious attention busy all the time, but what happens when we just stop and be present? We can let our awareness have no preferences or desires. We can accept whatever situation we believe we're participating in. For all we know, it may be an advanced virtual reality show for our participation in different levels of energy, so that we know the feelings of a wide range of low-vibrating emotions. We participate in its design and energy patterns all the time, and we design it to keep our consciousness focused on fear-based experiences by believing in our mortality and the reality of space and time. Now our challenge is to be able to

Chapter 5. Self-Realization

break out of the hypnotic trance of humanity and restore awareness of our true Being.

Reasoning with ourselves can work up to a point. We've conclusively proved that death of the body does not change our personal awareness. Everyone who has been dead and came back, and has reported, stated that their personal conscious awareness did not change, and they had memorable experiences and encounters while out of the body. It was also beyond time and space.

Now the challenge is to align our personal belief with the reality that we now know. Enter our ego, who is an entity created to deal with fear for our mortal survival and entertainment. The ego has no true heart connection and little intuitive guidance. It is empirically oriented and has its own selfish interest foremost in a world of struggle. It does not know that we are eternal personalized Beings of Creator consciousness. How can the ego believe in eternal Being? It is beyond comprehension and certainly beyond the experience of almost everyone we know. Is alignment with the truth possible?

When we begin acting from a perspective of love and compassion, that is at odds with the prevailing belief patterns, we can begin to feel truly free. We are breaking out of the trance. Once we expand our awareness beyond the empirical realm to the world of pure energy, we can be aware of our great conscious potential in all ways. We instinctively know how to direct energy patterns. We do it all the time, because we're designed to create in every moment with our thoughts and emotions, according to our perspective in life. The vibratory frequency of this energy determines the quality of energy that we attract into our experience. We can create whatever we want, when we have mastered our ego through love and compassion with the reward of relaxation.

By knowing our eternal essence of Being, we awaken fully to Self-Realization as the essence of the Creator. We participate in universal consciousness, knowing that we are all the same

Being, playing different personal roles for the enhancement of our experiences.

Knowing Our Eternal Being

How far can we expand our imagination? Mathematicians constantly strive to discover greater solutions to problems that were unsolvable. Movies have expanded the consciousness of humanity into greater self-destructiveness of super villains, as well as greater human high-vibration abilities, some of which have been shown in science fiction and super-hero stories. Humanity is just awakening to the idea that we are more than human. We are intent on pushing our limits higher.

Quantum physics has shown that everything that exists has an energy pattern unique to each conscious entity. It has also shown that there is a universal consciousness beyond time and space, that is the cause of everything and is the Creator of the quantum field of all potentialities. Our personal consciousness arises out of the consciousness of the Creator as a fractal of the Creator. Each of us is a Creator.

In the current human dimension, we have limited our creative power to the vibrations of a fear-based perspective. We have created everything we can imagine and feel, as well as our current circumstances. We are constantly creating with our thoughts and feelings. As long as we have fear of termination in ourselves on any level, we will have a fear-based perspective.

Only when we truly know and consciously experience our eternal Being, can we be fearless and truly loving of all. Perhaps we can be convinced by all the stories of those who died and came back, and who reported that nothing terminal or diminishing happened to their conscious self-awareness, and they became aware of much more. Because the quantum field, that they experienced outside of the body, is a field of unconditional love and vitality, most did not want to come back to their bodies, but did so out of a felt obligation.

By imagining ourselves to be eternal Beings, we can open ourselves to the truth of who we are. In our deepest Being, we have an intuitive knowing that we know is true about ourselves. This is where we know we are personal self-conscious Beings, who are eternal, created beyond time and space in the eternal universal consciousness of the Creator.

Living in the Fullness of our Being

We are designed to live in the brilliance of our inner light, enjoying unlimited freedom and abundance in all aspects of living. The essence of our Being is unconditional love in eternal presence of divinity. We constantly receive the living essence of the One universal consciousness flowing into us, enlivening us and providing everything that we modulate into being with our unlimited creative conscious awareness. The more gratitude we express in all aspects of life, the more beauty and joy we get to experience.

In our personal consciousness we arise out of the Being of the One universal living consciousness that creates universes. We are the same essence as the timeless One without limitations of any kind. How do we as humans not know this? Because we are unlimited, we decided to create limited, apparently separate expressions of ourselves in order to know what we could not know in our true being. We've learned what it's like to experience ourselves as victims and even despicable characters. We've made this human experience so convincingly real, that we've been completely unaware of who we really are.

Now we're ready to awaken from this hypnotic trance and return to awareness of our true Being. We've begun the journey inward to recognition of the expression of our divinity in the energy of our heart and the light of our soul. Our true Being is gently urging us to awaken to the love and joy of the presence of our Creator within each of us. We are becoming aware of our unlimited creative potential and are realizing that we have created ourselves to be needy, unhappy and unfulfilled. Now it is time to create ourselves out of our predicament and return to our expanded Selves.

There is much guidance available to us now, as all of humanity is awakening out of the human Matrix situation. We're leaving fear behind, just not paying attention to it anymore. We're ready to live in a society based in love and kindness, as we create the new world that we truly want to return to.

The Presence of the Master Within

Humans are naturally bound to ego consciousness. The ego is necessary to navigate our way in the world, but it is limited to the operations of our mind and emotions. As ego-conscious beings, we are vulnerable to every kind of mischief and limited understanding. We hold onto our connections with our family members and larger clan in our attempt to find solace and love and a meaningful life. None of this ultimately provides us with the deeper sense of being that we seek, as we all pass in and out of the body. It is only after we transition out of the body that most of us discover our true Being.

It is possible to know our eternal Being while confined to the physical body, and it requires a deeper penetration of our conscious awareness. It is helpful to align with the increasing resonance of the Earth and to open our awareness to the natural vibrations of our heart. As we invite the creative energy of the Source of our life stream to clear our pineal gland and awaken our awareness to our larger Being, we begin to feel the presence of universal consciousness enveloping us in the quantum field of infinite patterns of light and unconditional love.

This usually requires a prolonged focus of attention and strong intention to know the truth of our Being. It can be practiced in our every-day interactions in life. By continually looking within to know the guidance of our intuition flowing through our heart, we transform our lives in alignment with the consciousness of our Creator. We become the presence of the Master within.

Living Intentionally

We can open ourselves to experiences of great beauty and abundance in every way. We can feel the greatest love and joy. This is our natural state of being. The only limitation keeping us from high-vibrational living is our willingness to accept living in unconditional love, while enjoying infinite creativity. Why would we resist our natural state of being, when we know what it is and how it feels? Perhaps we just don't believe it could be true. There are a lot of contra-indicators. It's outside the spectrum of energy that humanity lives in, so even to be able to imagine being in absolutely wonderful experiences, is a great stretch for us. It is a mental and emotional rise in vibrational dimensions.

We must confront every possible fear in our emotional bodies, including fear of termination of consciousness. This is what a vision quest is all about. It is dropping everything that is not important for survival and going to a remote place where we can consciously meet our essential Self, our present awareness of Being. We may expand our awareness beyond the spectrum of vibration of normal human awareness. We begin to see brighter colors in greater numbers, we hear colors and see sound and rainbows unending. We begin to enter universal consciousness, overwhelmed with unconditional love.

It is not necessary to follow the traditional path of the vision quest, but we do need to feel within that we are part of everyone and everything. Learning to feel the energy of the Sun, the Earth, the trees, the birds, the nature spirits, the water, the air and even fire and volcanism, are all part of universal consciousness. We can become aware of etheric energy patterns, which are the manifesting wave/particles that we recognize into empirical form. We become conscious creators in our larger Selves.

We can realize our unity with our creative Source. Once we have released all of our limiting beliefs about ourselves and realized the vibrations that we love the most, we naturally attract experiences that resonate with high-vibrations, and the realm of fear disappears. Only love remains, together with great vitality and exuberance for life. The traditional spectrum of human energy becomes for us a realm of experience that no longer holds fearfulness or unfulfilled needs of any kind. We become the masters of our lives and all of our experiences through the intuitive knowing of our heart.

Journey to the Inner Light

Within each of us is the presence of the eternal One conscious Being, Who constantly creates us personally. There is nothing permanent, except universal consciousness and the field of energy that emanates from it, constantly changing patterns of electromagnetic waves of energy. Some of these energetic patterns vibrate within the spectrum recognized by humanity, but our inner knowing is aware of a much greater range of frequencies that reach into a higher dimension of consciousness. We can have this expanded awareness only if we intentionally open ourselves to our eternal presence of Self.

We can resolve all beliefs in our limitations, as they arise, and as we can keep intending to know our true Being and the essence of our consciousness. We can feel all the vibrational patterns enveloping us, and we always know when we're in the presence of low-frequency fear or high-frequency love. We can keep intending to align with love and joy, whenever possible. We can align with nature and the vibratory resonance of our planet.

We can take the journey inward to the essence of our Being. We can feel the quality of the vibrations of our heart. We find them by intending to recognize them emotionally. We know what love feels like, even though the love we can know as humans is limited. If we can go into its depth, we can feel the unconditional love radiating from our heart, who lives to give us life. Through our heart, we constantly receive the unconditionally loving conscious life force of the Creator.

We know what it is to feel alive and aware. Part of our aliveness and awareness resides in our body consciousness, but there is much more. There is beauty, joy, abundance, music and magic. It is all possible for us in a perspective of love, if this is what we search for. We can open ourselves to a higher dimension of living by creating it in our imagination and emotions, until we can recognize its reality. The higher dimension of living exists for us as soon as we know that it exists. It happens when our vibrations resonate in alignment. This is all part of our inner journey. We can learn to be aware of our inner promptings, guiding us in every moment to the full realization of our magnificent eternal Being.

Exploring Our Consciousness

If we want to know our true Being, who we really are, and our capabilities, we must intend to explore the depths of our conscious awareness. We are individual persons with our own self-awareness. As the human spectrum of resonant frequencies comes into our awareness, we need to make a jump in our vibrations to be able to be aware of the vibrations of our heart. We must shift our emotions into the spectrum of love and joy, our natural, created state of Being.

In our Being beyond the physical, we can have the perspective of our true Selves. We are beyond time and space in our essence of Being. We can feel the flow of life force arising for us within the universal consciousness of the Creator. In unconditional love we are all the same Being in our essence. We are Creators within the Creative consciousness of everything. We are constantly being created to create our own experiences within the universal consciousness by modulating the energy patterns that we encounter with our own visions and feelings in our focus in every moment within the limitations of our beliefs about ourselves.

All beliefs are creations of our ego in its attempts to make life tolerable without needing too much understanding. They have guided our ego in maintaining consistent energy patterns that we are familiar and comfortable with. Taking on a higher-frequency perspective requires strong intentional focus on the love vibration. The challenge with this attempt is that without proof, in the perspective of the ego, we must intentionally open ourselves to the unknown portion of ourselves.

An intuitive sensitivity guides us in our inner journey. It is what we deeply know without words. We can seek the level of vibrations that feels like a sustained state of gratitude and joyous ecstasy. The conscious boundaries that can keep us from knowing this are our beliefs of what is possible for us. In our essence we are eternally-created, limitless self-conscious Person-Creators. We only need to realize ourselves beyond limits. We can do this while also participating in society. We just become more radiant.

Awareness of Our Energy Spectrum

There is no competition for enlightenment. Eventually everyone will get there by many circuitous routes. We always have free choice for our attention and focus; however, we've been trained

Chapter 5. Self-Realization

and programmed to believe that we must vibrate within a socially-acceptable spectrum of energy. Those beliefs are all based in fear of anticipated suffering and termination. Apart from our alignment with their energy patterns, they could not exist. Fear is not a real thing. We imagine it into our recognized reality. Our life force goes out through our attention to provide the vibratory pattern that we radiate into the quantum field, from which it becomes recognizable for us as real. By withdrawing our attention and refocusing, we stop creating the old frequencies.

We have the ability to pay attention for our creations. We can select the vibratory frequency that we choose. We can feel and imagine living in that spectrum of energy. We can choose to experience vibrations that are beyond the boundaries of the empirical world, frequency patterns such as truth and clarity of consciousness. We can choose to know and feel the true energy of our heart and our intuitive knowing. We can know when we are focusing in our true, natural energy patterns in a higher dimension. This is the spectrum of joy and everything that aligns with it.

We are so much more than our human selves. We do not need our physical bodies, except for our empirical experiences. If we take our attention away from the body and focus on just being present in awareness of our Being, we can feel and know in our heart and intuition that we are eternal self-aware persons.

Our natural, most fulfilling vibratory frequencies are in alignment with the rising resonant frequencies of Gaia, Spirit of the Earth. We can intend to feel and align with these vibrations, which also align with our heart, by being in nature, especially in beautiful and spectacular environments, and being aware of the qualities of energy patterns we encounter, as well as the vibrations that come through our bare feet on the Earth.

Our Creator consciousness expresses itself as our personal energy signature in the quantum field. The energy patterns that we create in our own consciousness radiate their vibrations into the quantum field of all potentialities, attracting energetic patterns that resonate with us. These become our experiences. We can have control of our lives just through the focus of our attention. We can elevate our lives by our awareness and intentional alignment with our intuitive knowing.

Resolving Our Limitations

Prior to mastering life in this dimension, we must resolve all emotional knots and limiting beliefs. We've acquired them through contracts that we made prior to incarnation, together with coping with difficult and potentially lethal situations in this lifetime. Our limitations keep us locked within the spectrum of fear. We could never escape the prospect of termination, which, humans believe, is going to end our conscious existence. Some of us have taken enough consciousness-altering plants, or meditated deeply enough, or have witnessed a number of peaceful death-bed transitions, that we know that we continue in our consciousness, regardless of whether we have a body or not. Belief in mortality beyond the body is the primary limitation that we must resolve in order to live beyond fear.

Watching videos of people who have come back from the dead and reading books of their accounts can be helpful in penetrating the truth of our Being. Studying quantum physics experiments may be helpful, especially those that led physicists to conclude that consciousness is the source of everything, and there is one consciousness that is universal. We participate in universal consciousness as much as we are willing to realize. Full realization comes by being our eternal presence of awareness.

The process of awakening to our true personal identity beyond our present awareness, can be assisted by convincing our ego that our consciousness exists beyond time and space and also includes time and space in the empirical dimension. It is in this vibratory spectrum that mortality appears. Everything here dissolves and is created trillions of times a second out of the unified quantum field of universal consciousness. In the empirical spectrum, this process appears to stop at the death of the body.

The energy patterns of our consciousness do not disappear, they change into a different dimension of frequencies, and the patterns are more ethereal and more easily modulated by our creativity. Here mortality does not exist. We recognize our eternal Self, our consciousness expanded beyond time and space. We realize that we are unlimited in our awareness and creative ability. We are the Creator, and the Creator consciousness experiences through us. We are designed to create experiences. This is why it is important to imagine and feel grateful, loving and compassionate in every interaction and confrontation. Doing this in our current world may get us battered for a while, but we can live in a world of high vibrations by creating it in our present awareness. This attracts compatible energy patterns that become our experiences.

Being in Conscious Unity with All of Life

We were created to be able to experience consciously the entire spectrum of energies in all dimensions and to continuously create energy patterns with our thoughts and emotions. We are the creators of our thought patterns and emotional feelings in every moment. These two intersecting planes of electromagnetic wave patterns generate an energetic signature that emits photons throughout our human spectrum of energies within the plasma field of universal consciousness. Out of this consciousness arise wave patterns that align with our energetic signature, and which become our experiences in whatever dimension we inhabit.

We are constantly having experiences interacting with many different kinds of conscious beings, all of us acting freely however we choose. In the lower vibrations, we choose the life of enslavement to fear, ultimately fear of the termination of our consciousness. In this realm we have been blinded to knowing our greater Self, the One through Whom our life force flows to us. We can open ourselves to feeling this energy flow. By being aware of the source of our conscious Being, we can be beyond fear. We can realize our eternally present Self-Awareness.

Our true Self Awareness is unlimited in every way. We have an open consciousness that includes everything in universal consciousness that is present in the quantum now moment, which is beyond time and space. We can freely focus our attention on any quality of vibration that we desire to feel and envision. This is our nature. By directing our attention to higher and higher vibrations in energy patterns, we can learn to react and be compassionate, loving and understanding in all encounters, regardless of the vibrational pattern that we face.

When we can maintain a state of being that is completely objective, our emotional and mental planes will intersect at a zero point of neutrality in resonance. At these frequencies of wave patterns, we can be at rest mentally and emotionally or engaging in fulfilling interactions. By maintaining high frequencies in ourselves, we elevate the energy around us, and the patterns that cannot align with us begin to disappear from our experience. It is only when we allow ourselves to align with low-vibration, fear-based energy, that we create unenjoyable experiences for ourselves.

By aligning ourselves with the Source of our life force and consciousness, we enter an awareness of the greater Self of all creatures, who are all in our conscious Being, where we know unconditional love and a deep feeling of connection to all conscious Beings.

6.

Universal Consciousness and the Quantum Field

The Nature of Our Reality

Quantum physicists have found that everything in the empirical world appears and disappears continually every micro-second, perhaps trillions of times per second. Our bodies and everything we experience is here and not here all the time. What is the source of all of it? How does it all come to be all the time? How do some things remain constant and others change minutely or greatly?

When physics experiments demonstrated that sub-atomic particles display consciousness on a cosmic scale, scientists finally were forced to recognize consciousness as the basis of everything. All electro-magnetic wave patterns and the appearance of recognized particles are designed and created by consciousness. All wave patterns and material forms are conscious beings. This consciousness is universal in the sense that everything and everyone in existence participates in it. It is the self-consciousness of us all, connected to everything in the same Being, who experiences through us everything that we think and feel and experience. We are that Being. Every aspect and ability of the universal Creator Being is in our consciousness, and it is available for us to recognize and to make it real for us. We can have any experience that we can imagine and feel, and we can even go beyond our limited imagination, once we resolve all of our inner limitations and beliefs.

By our participation in the living spectrum of the energy signature of humanity, we have developed all the beliefs about ourselves that we needed to stay locked into this limited realm of energy that we recognize as material. It is our recognition and belief in the reality of this realm that makes it so. Now we can move beyond these limitations by wanting to live wonderful lives of unlimited beauty and joy.

All we need to make the leap in consciousness is a micro-second, once we are fully present and intuitively receptive to the energy of our pure heart. This is where we realize our universal consciousness in the quantum field of all potentialities available for us to create. This is where we regenerate our bodies and create our experiences. In being fully present in our conscious awareness in the now moment, we can feel our life force streaming into our Being in unconditional love and joy. Our conscious awareness expands to infinity. There are no limits to our creative ability in every moment.

Our Cosmic Consciousness

Quantum physics has shown that there is a universal consciousness that provides the energy that comprises everything that exists and the operating awareness of everything that moves and has being on the subatomic scale, and by extension every-

Chapter 6. Universal Consciousness and the Quantum Field

thing that is composed of atoms and their components. Without universal consciousness, there would be nothing in existence. Everything participates in universal consciousness. It is the essence of our being. Physics experiments have shown that sub-atomic particles have cosmic consciousness. They know and act with immediate certitude when observed and when moving through obstacles with available pathways. They are quantum in their being, which means they can be in more than one place at the same time. They are not limited to time and space, which are prime requirements for the empirical world. They are both empirical and non-empirical, depending upon whether we recognize them. What does this have to do with us?

Once I wrote that the whole (meaning us in our embodiment) cannot be less than its parts. My implication was that, if our constituent atoms and their subatomic entities have cosmic awareness and inter-dimensional abilities, we must also have these capabilities. The human masters of consciousness (Jesus, for example) have demonstrated that they do have them, but what about the rest of us? If we don't currently have these abilities, can we develop them?

It all relates to our free will and our understanding of why we are living on the Earth. Our awareness of our essence, our self-conscious presence, is not limited to our bodies, but most of us identify ourselves with our limited ego consciousness, even though our bodies have much greater awareness than most of us recognize, and that is the reason we are limited. We recognize ourselves as limited.

Life on this planet is a training program for our consciousness. We are here to experience what we could never imagine in our true Being. Collectively we have designed and manifested a dimension of consciousness that keeps us from knowing who we are, so that we can explore the dark side of life, the kind of consciousness that is destructive and fearful. It is so that we can expand our consciousness into a greater awareness of what is possible, so that we will be able to have greater appreciation for the unconditional love and joy that is our essence and the essence of all that exists. This experience will be useful for us in our future creation of universes. We have all the capabilities of our Creator and the universal consciousness that we in our essence continuously arise out of and experience in our true Selves. We can learn to train ourselves to recognize and identify with who we really are, and this is the movement we are currently becoming aware of.

To Be the One

What is the nature of the quantum field of all potentialities? It is invisible to us and cannot be entirely perceived by our instruments or our senses. It must be known intuitively. We know of its existence through quantum physics experiments, in which the behavior and abilities of sub-atomic waves/particles have given us clear evidence of their conscious awareness and ability to interact with us. Their interactions happen instantaneously everywhere that we could detect. These are abilities impossible for us to imagine in terms of our technological knowledge and traditional physics. All interactions happen outside of time and space. They happen in the quantum field. It is a field of universal consciousness and life force with the ability to create anything and anyone. It knows everything always and everywhere. It has no boundaries and is infinite in every way. It is the expression of the Creator Being, the Source of all consciousness and life force.

Everything that exists is constantly arising from the quantum field and demising back into it, including our empirical selves. Our true Being always exists in the quantum field. We are eternally present in conscious awareness. We can manifest our presence in any way that we choose. We decided to create a compartment in our consciousness for our Earth experience, and we have believed ourselves to be encapsulated in this portion of our Being.

Chapter 6. Universal Consciousness and the Quantum Field

We have been unable to feel and know unconditional love, because of our encapsulating social training. Once we know unconditional love, we cannot have fear. It disappears, because its frequencies cannot align in resonance with love, which is the life force. Fear has no life force and needs to be created by conscious beings.

Whenever we are ready, we can awaken to our true Being. We only need to resolve all of our conscious boundaries that keep us from recognizing the feelings and vitality coming through the energy of our heart. These are all high-vibration emotions that feel wonderful and compassionate. If we stop resisting unconditional love and instead welcome it into our being, we will naturally feel our connection with our true Being.

Jesus as a Quantum Physicist

Quantum physics has shown that we are multidimensional beings. We can be in more than one place at the same time. We can be aware of more than one dimension at the same time. We are all connected in our essential being, arising out of the quantum field of all potentialities. This is the nature of our Father/Mother God, Prime Creator. In our deepest Being, we are aware of the universal consciousness that enforms our being and everything that exists. Because of our unfettered ability to be aware of our presence and make all decisions about all of our experiences, we are limited only be our beliefs about ourselves.

Who do we believe that we are? Since humans are conditioned by allowing ourselves to believe in the limitations that humanity recognizes, we are creating those limitations for ourselves. We are taught to believe that we are limited by mortality, aging, having needs and being capable of being fearful. This is all an imaginary drama held as reality in the combined consciousness of humanity. Quantum physicists have seen the basic proof that extends to the entire cosmos. By recognizing and accepting our empirical realm as real, we constantly draw it out of the quantum field and continue creating it. It is real, along with its inherent limitations, in the realization of each of us.

Who are we really? Let's look to Jesus and other persons who showed similar abilities that humanity has recognized as masters. Jesus showed that he could be in more than one place at the same time. He could change the essence of material things. Healing diseases and disabilities requires the realization and belief of the reality of the healing on the part of the patient. Calling back a departed spirit requires the desire and belief of the spirit, because we are all sovereign beings, departed or embodied. Recognizing and communicating with the spirit in another dimension shows multidimensionality of being. These are all true and real in the quantum realm. Jesus showed us the perspective of the master, who knows the nature of reality beyond the material world. And he told us that we can also do everything that he did. Why don't we? What is the quantum secret?

It is the frequency of love and joy, realized within so thoroughly that we remember who we really are as master creators of universes. This is the vibration of the higher dimension and the expression coming from the quantum field. We only need to recognize it and allow it to be real for us.

Our Current Life's Purpose

We are involved in transformation and conscious expansion. We are intending to be in gratitude and joy as much as possible, and to connect with our divine essence. This is our primary process now. As we express high-vibration feelings and thoughts, we raise the vibrations of our energy signatures, and we become more radiant and influential. We draw each other into our awareness with love. We are the family of Light, and we are returning to our essence.

We are recognizing that we are multidimensional. We can live in a realm of high-vibrations while also being present in awareness of the drama of the human energy signature. With our radiance, we draw the frequency of humanity into our realm. Our creative love is more powerful energetically than any of the lower vibrations. Everyone knows the feelings of love and joy intuitively. These are the most attractive vibrations and the ones we feel best expressing. Having resolved our interfering personal beliefs about ourselves, we are free to be naturally unlimited in our creative ability. We can dance in the Light and create miracles.

Our emotional energy vibration is most important. This carries us into high-vibration experiences and states of being, in which we can expand into even higher-vibration experiences and miraculous scenarios. We can go as deep and as far in our awareness as we are willing and able to recognize. We can live in a world of love and beauty, whenever we allow ourselves.

We may need a thorough reorientation in our understanding of the nature of reality. Nothing except the quantum field of energy exists outside of our awareness. Whatever we recognize also is aware of us and becomes real in our experience. It all exists in our consciousness. We are the creators through our ability to modulate the energy patterns in the quantum field.

Our Eternal Essence

We continuously arise from the One Self-Conscious, Self-Realized Being that is all that exists. In quantum physics this consciousness expresses Itself as the quantum field of all potentialities. Every possible being and circumstance exists as a potential scenario within the electromagnetic wave patterns of the quantum field. Everyone and everything that exists is a conscious entity continuously arising and demising every microsecond.

How do any of these potentials get actualized? They get recognized by a self-conscious being and appear in their recognized form in the experience of that being. This happens through creative imagination and focus upon repeating experiences.

Upon incarnation, we lose awareness of our true Being, so that we can experience a life of apparently real personal separation from the quantum field of limitless possibilities for our experiences. We participate in a limited spectrum of energy that our consciousness has been programmed to recognize for manifestation into our lives. This is the empirical world. Although humanity recognizes this realm as real, it is actually imaginary, just like everything else, except our Being. All humans telepathically align with the human energy signature in order to live in this empirical realm.

We've intentionally become accustomed to limiting our awareness to the empirical spectrum of energy. We found that we had to severely limit the creative inflow of life force from the quantum field in order to make our experience more real. With limited life force, our egos have become entrained in low-vibration energies here and must be released into sovereignty and freedom, and intentionally living in our natural spectrum of vibrations in a higher dimension of experiences.

We release our ego from its limited awareness by loving it unconditionally. It has allowed us to live in this imaginary world and experience a spectrum of energy that would not otherwise have been possible for us. It has deepened our compassionate understanding. This is a really successful Earth experience. We can be grateful to our ego selves for all of this. Now the ego can an extension of our true personal Being, living in the awareness of our existence, constantly arising in the life force, unconditional love and eternal Being of the quantum field, the expression of the Infinite One.

Chapter 6. Universal Consciousness and the Quantum Field

The Nature of Our Reality

Quantum physicists have found that everything in the empirical world appears and disappears continually every micro-second, perhaps trillions of times per second. Our bodies and everything we experience is here and not here all the time. What is the source of all of it? How does it all come to be all the time? How do some things remain constant and others change minutely or greatly?

When physics experiments demonstrated that sub-atomic particles display consciousness on a cosmic scale, scientists finally were forced to recognize consciousness as the basis of everything. All electro-magnetic wave patterns and the appearance of recognized particles are designed and created by consciousness. All wave patterns and material forms are conscious beings. This consciousness is universal in the sense that everything and everyone in existence participates in it. It is the self-consciousness of us all, connected to everything in the same Being, who experiences through us everything that we think and feel and experience. We are that Being. Every aspect and ability of the universal Creator Being is in our consciousness, and it is available for us to recognize and to make it real for us. We can have any experience that we can imagine and feel, and we can even go beyond our limited imagination, once we resolve all of our inner limitations and beliefs.

By our participation in the living spectrum of the energy signature of humanity, we have developed all the beliefs about ourselves that we needed to stay locked into this limited realm of energy that we recognize as material. It is our recognition and belief in the reality of this realm that makes it so. Now we can move beyond these limitations by wanting to live wonderful lives of unlimited beauty and joy.

All we need to make the leap in consciousness is a micro-second, once we are fully present and intuitively receptive to the energy of our pure heart. This is where we realize our universal consciousness in the quantum field of all potentialities available for us to create. This is where we regenerate our bodies and create our experiences. In being fully present in our conscious awareness in the now moment, we can feel our life force streaming into our Being in unconditional love and joy. Our conscious awareness expands to infinity. There are no limits to our creative ability in every moment.

Humanity's Great Delusion

We've all had to accept the limited consciousness of humanity in order to live here in this dimension. Humanity lives in a frequency spectrum of electromagnetic waves that we perceive as the physical world. It seems solid to us, and it stimulates all of our senses. But is this all there is? How can we know from within our limited consciousness?

We can expand our consciousness intentionally. The findings of quantum physics may be helpful. We know through scientific experiments that time and space do not exist apart from the conscious fixation of humans. There is no physical world, apart from the recognition of humanity.

If humans did not consciously observe the wave patterns of humanity's energy signature, the empirical world would not exist for us. It would just be patterns of energy waves, until we observe this spectrum again. Then our observation of these wave patterns manifests materiality in our experience instantly, because we are creating it with our recognition.

This exclusive constant focus within humanity's energy spectrum makes us appear as if we are exclusively material beings, with no awareness of vibrations beyond humanity's spectrum. We have deeply imprinted limitations on our perceptions. These we must resolve with compassion and love.

Our recognition creates our awareness. If we can recognize higher vibration scenarios, we experience life at those levels of vibration. We are unified beings, and our emotions generally operate in resonance with our thoughts and imaginings. To recognize the essence of something, we must be aware of its energy signature, which we feel emotionally. It is our sense of the quality of presence of being.

Every being is sovereign and eternal. Everything is conscious, from the smallest sub-atomic entities to the galaxies and beyond. Consciousness is unlimited. It is everywhere in all dimensions of timelessness. Consciousness is the source of everything and everyone. It expresses itself as a unified quantum field of all potentialities. This is a field of plasma that is aware and responds to consciousness always. It provides the life force for every conscious entity and connects all through itself. It experiences everything every entity everywhere experiences.

We arise out of the unified quantum field with our own unique energy signature. It is our presence of being. Our essence has the ability to resonate the energy patterns that we encounter. With our emotional power and our imagination, we can create scenarios that will attract resonant beings and situations. By being in joy and recognizing it, we attract joyful situations and beings, and we contribute to raising the energy signature of humanity. We participate in universal consciousness, once we have resolved all of our self-imposed limitations.

Rising Human Destiny

Now that quantum physics has identified universal consciousness as reality, we can know that our potential awareness is universal, and that all conscious beings are intimately connected in the same field of consciousness. In essence we are every conscious being, and we can intuitively and telepathically connect with all beings everywhere. That includes snails and whales and everyone in between. Some of us have achieved this, as we awaken to our true Being. The destiny of humanity is Being our true Selves.

Humanity currently holds the vibratory spectrum of the human matrix in our collective consciousness. We are constantly creating it by modulating the energy patterns that we encounter with our own attention, within the limits of our beliefs and perspectives, which are all based on fear of some Kind.

We are getting closer to our essential Being, whom most humans cannot yet even imagine, having been deeply misled. By focusing on high-vibration scenarios with our thoughts and feelings, we can participate in a higher quality of life in every way. We are beings of light. We emit radiant biophotons comprising our aura and presenting our presence of being in the quantum field. Our aura is the radiance of our energy signature.

Life in the human matrix on this planet is coming to an end in its current vibratory level. As the resonant frequency of the Earth rises, as it is doing now, all conscious life on this planet must be in alignment. If any conscious entity does not raise its frequency in resonance with the Earth, its vibratory frequency will become unstable and dissolve.

Every human has a choice of what vibratory level to live on. Without awakening to true Being, that choice will be very limited because of humanity's beliefs about our identity. In our own consciousness we can resolve these limiting beliefs through a perspective of compassionate wisdom. Our ego can relax and enjoy being an expression of an infinite and eternal Being.

The species consciousness of humanity by design is destined to create a higher dimension of reality. This is a natural expression of the jump in vibration of humanity's energy signature that will align with the increasing vibratory frequency of the Earth. The most powerful life current that is expressed through the quantum field is rising in frequency toward our realization of unconditional love and joy. We are allowing the low-vibration realm to disappear, as we withdraw our life force through our attention, which we can give to high-vibration feelings and awareness.

Each of Us Has Our Own Reality

Quantum Physics has shown that sub-atomic particles are particles only when we recognize them. If we're not recognizing them with our senses or our technology, they immediately manifest as electromagnetic waves with their own unique frequency and amplitude. They reappear as particles as soon as we focus on them again. This is a conscious process on their part. Their alignment appears to extend throughout our known universe. They become part of our physical body by vibratory attraction.

Because our physical body is a collection of atoms and molecules, each of these with its own consciousness, our body needs to be entirely in resonance in order to be healthy and well. If there are serious anomalies in our force field, they can ultimately destabilize our entire physical manifestation and cause dis-ease and diminishment of our energy signature.

To be truly in the power of our natural life force, we need to be clear emotionally and mentally. Until we attain this state of being, we can continue to resolve, through love and compassionate wisdom, all issues that arise. We can be intentionally open, in our feelings and imagination, to inspiration from the intuition of our heart. Here is our higher guidance, if we can understand it. One way to do that is to be deeply sensitive to our feelings about everything we encounter in our experiences or in our imagination. Higher guidance is always immediate and usually subtle and symbolic. We can feel the quality of the energy, its vibrations, and we know when we feel expansive and joyful.

If we focus on the emotional quality of the energy of our heart, we will always be fearless and compassionate. This is high-vibration energy. It attracts energetic alignment throughout our body, down to the atoms and molecules. All will vibrate in resonance, and we will feel wonderful, in our natural state of Being.

With the increasing vibrational resonance of the Earth, all consciousness on this planet is being attracted to higher-vibration living. In order to continue to live here, humanity is raising the frequency of its energy signature through increasing awareness of joy and peace. This is happening because more and more of us are realizing that the game of 3D is ending, and we intend to enter a higher dimension of living.

What Do We Really Know About Ourselves?

Quantum physicists have discovered that everything participates in conscious awareness, beginning with all sub-atomic waves/particles. Everything that we know about exists in the nature of energy patterns that become material when we observe them and recognize what they are. Through our observation and recognition we create our material reality. There exists a realm or plasma-filled space of electromagnetic waves and patterns that can be recognized as anything we can imagine. This realm is known as the unified quantum field of all potentialities or the vacuum fluctuation with constantly flowing and changing patterns of waves of energy. This realm also contains the universal life force that enlivens all conscious entities and endows them with self-realization and unlimited awareness. Photons, for example, have been observed to know instantly the condition and operation of all photons everywhere that physicists could observe them. By extension we can know that we also have this universal consciousness, limited only by our beliefs about ourselves. This has been confirmed by many who have experienced out-of-body explorations.

We have been living as if we are separate beings without a connection to universal consciousness and life force. We have been taught that we are nothing more than mortal beings, subject to many kinds of threats and suffering. Religions have offered us respite in an afterlife of reward or punishment, provided we believe what they tell us. How do we know what is true about our nature?

Quantum physics has implied that everything we experience is a reflection within our own consciousness. Because we have universal awareness, we can be aware of the awareness of all beings with only the limitations that we impose or allow to be imposed upon us. These limitations have kept our awareness locked into a Matrix of energetic patterns of low vibrations of anomalous, incoherent energy. We have believed that we can get sick, be disabled, suffer and die. But we can penetrate our innermost being to learn the truth of our Being and ultimately become aware that we can participate in consciousness encompassing the entire cosmos.

Many masters throughout history have taught us how to be aware of the intuitive guidance and knowing of our heart, how to build up our life force, to regenerate our bodies, to be aware of our eternal Being. We are not mortal or dependent upon anyone or anything for our well-being and enjoyment of life. We have the innate ability to create whatever we want in our perspective of authentic Self-realization. We are embodiments of the intelligence and Being of the Creator of all that exists. We can live in love and joy as we use our creative abilities to support and enhance all of life. In our essence, we are all the same Being, each of us having our own special talents to create experiences for ourselves and the Creator, within whose consciousness we dwell forever.

Our Emotional Essence

All Being emanates from the consciousness of the prime Creator, expressed in the quantum field of all potentialities. Out of this we arise as personal Beings, having all of the abilities of the Creator's essence. We are constantly created to have experiences of all kinds for the Creator. We are the Creator in the same way that our fingers are part of our body. We experience things through everything that we create with our attention, imagination and emotions. Our attention arises from the essence of awareness of our Being. Our imagination is our innate ability to modulate patterns of energy and create what we want or don't want. It all depends upon the vibrational focus of our attention, the vibrational quality of energy we pay attention to. Is it of love or of fear? Our emotions tell us what quality it is. They are part of our divine consciousness. They are our energy transmitters, our creative power and our awareness of the vibratory frequency that we focus on.

Emotions arise from the unlimited consciousness of the Creator and flow through us with our life force. When we are silent mentally and emotionally, we can be aware of the natural flow of life, which we can feel and know intuitively. We can focus our attention far beyond the limitations of our beliefs, and our emotions can carry us there. They'll come up against the blocks in our consciousness that inhibit expansion. These we can resolve with our awareness in a perspective of compassionate wisdom. We can and must resolve any limits that we encounter in recognition, gratitude, forgiveness and love.

We can resolve the attachments created by the ego, our false beliefs about ourselves and everything that is part of the realm of fear and mortality. We can turn our awareness inward to feelings of joy, abundance, peace and all of the high-vibrational emotions. They will carry our awareness into the natural flow of our life and guide us in our intuition. We can be open to living in the flow of unconditional love in our life force, which connects us with all beings consciously. This is part of our conscious expansion, ultimately into infinite awareness.

Living in Eternal Being

What really is our potential in this life? We can live in great adventures without fear, while knowing our eternal Being beyond physical and ego consciousness. We are constantly arising out of the conscious Being of the Creator of all. This is the consciousness that fills the unified quantum field with a plasma of unlimited numbers of electromagnetic energy patterns and clouds of photons. From the quantum field arises everything that exists in form and experience. Every entity has its own consciousness and shares this in the universal consciousness enveloping the cosmos. This is the creative expression of the prime Creator. Every thought, emotion and experience is carried in this consciousness and are the experiences of the Creator. We are the essence and experience of the Creator. The Creator is our Being. This is who we are. We are created to modify and modulate the energy patterns in the quantum field to enhance the experiences of the Creator, who creates through the consciousness that we are. Once we have this realization, we are in a completely expanded sense of awareness. We know things that we cannot know in the limited consciousness of humans. We feel emotions that vibrate beyond our current range of awareness.

All of this is in our consciousness, but we've closed ourselves off to it in order to have the full human experience in physicality. We can be in the process of resolving and eliminating all of our limitations, which we convinced ourselves to enclose our reality. We are actually living in an augmented reality that we cleverly designed eons ago. It's all based on fear in all its possibilities. There is no where in this spectrum of vibrations that allows for our awareness of our eternal Being. We've been taught that this is the only spectrum of energy, the only reality, and it is so for us, as long as we believe it. Our beliefs create our limitations.

One path to awakening to our true Being is just to be open to the rising vibrations of the Earth and our cosmic environment.

This entails being sensitive to our intuition, which prompts us with feelings, inspiration, words, images and many other forms of communication. These are all intended to guide us in every moment toward our freedom and sovereignty and resonant actions. Our emotions tell us the quality of energy in our awareness, and they can also guide us with our intuition. Much depends upon what we desire to be aware of. Everything we could desire is within our own consciousness. We can learn how to open our awareness to it through aligning ourselves with rising natural energy patterns.

The Meaning of Life

We dwell in a realm that we recognize as empirical, because we perceive it with our physical senses, but what do we really perceive? Quantum physicists have shown that we live in a sea of energy patterns that become physical when we recognize them. We are constantly creating everything that we recognize. Everything arises from our consciousness.

We have been trained to believe that our consciousness is separate from all others. Let's consider the nature of consciousness. It is our awareness of our personal identity within the bounds of our beliefs about ourselves and our awareness of everything we perceive, imagine and emotionally feel. It is our knowing of who and what we are and the environment that we inhabit. All limitations to our awareness are self-imposed as beliefs and recognitions. Everything that we perceive is a reflection of our inner awareness. We do not perceive anything outside the parameters of our beliefs. These are the boundaries of our conscious awareness.

What we're doing here in nature and in human society is having experiences in our awareness. How do we know what our awareness is capable of, if we can't be aware of anything more? We have to resolve our beliefs and fears, so that we can rely on our intuition, which is where we know everything. We can align ourselves with the vibratory frequency of our innermost Being, from which our intuition arises. This is the high-vibratory spectrum of joy, inner peace, beauty, gratitude and love. Our intuition can draw us into alignment, because this is where we receive our conscious life force. As we come into vibratory alignment, we become aware of the unconditional love that envelops us and connects us with all other beings in the universal consciousness of the essence of the Creator, of whom we are each a fractal, essentially a duplicate Creator, able to be and do anything we desire. We can envision living in beauty, pursuing everything we most enjoy. Our interactions with other beings are expressions of heart-felt joy and kindness. We can be aware of being eternally present now, while also participating in life with others who do not necessarily have any desire to open their awareness and raise the quality of their lives. We can be present and fully aware in every moment and in alignment with our intuition.

Guiding our Chosen Destiny

Quantum physicists have determined that the origin of everything is the universal consciousness or primal cause, that expresses itself as the unified quantum field of all potentialities or the vacuum fluctuation. This is the realm of electromagnetic energy waves and patterns. It is unrealized energy acting and reacting in every possible way. In spiritual conceptualizations and most religions, this universal consciousness is called God, the Supreme Creator. The Universal Consciousness creates conscious Beings, who are extensions of Itself, fractals, which are complete expressions of Itself. All conscious Beings arise out of the Universal Consciousness and are universally conscious. Humans receive from the Universal Consciousness the ability to modulate the energy waves and patterns in the quantum field. This makes us infinitely powerful creators.

If we imagine it, we create it. These words are true, but most of us have not been aware that it applies to all of our mental and emotional processes. Our energy signature is constantly changing with our thoughts and feelings. We can put the ego to rest only when we are completely fearless. We must be vibrating in love.

How can we, as humans in our current spectrum of vibrations under the limitations of our (false) deeply-held beliefs about ourselves, possibly believe that we could be creators of universes and more? It's a big stretch of our imagination and emotions. It asks us to experience personal immortality in our awareness. We may have to construct scenarios that will take us in that direction. Our ego consciousness needs to be put to rest. We can love and be grateful to our ego for getting us to where we are. We're still alive in the body, and now the ego can be content and enjoy watching the show. We'll call on it when we need to do things. We now have higher guidance in our intuition. Universal Consciousness is our real home. We've just deployed part of our consciousness to participate in this human experience. Now we're being called back to awareness of our whole Being.

We can be aware of our vibratory frequency through our emotions. If we intend to go higher, we can learn to be more loving, more compassionate, more joyful and more confident in our eternal presence of Self-awareness. In creating our new world, conscious action may be called for, but the real creative power is our consciousness.

Our Semi-Real World

Our human life experience provides us with the opportunity to transgress universal laws and to engage in energies that are not possible for us in our true Being. We've been able to exercise

Chapter 6. Universal Consciousness and the Quantum Field

tremendous power over others, suffer poverty, enslavement and torture and feel unscrupulousness, fear, hatred, jealousy, envy and lust that are all low-vibration emotions beyond our imagination as eternal, infinitely powerful creator Beings in our true essence. We came here to experience these things in order to know deeper compassion and greater love in our expanding consciousness.

In order to experience low-vibration lives, we had to compartmentalize our consciousness by creating powerful limiting beliefs about ourselves and our abilities. We erased our memory of who we are as we incarnated into an artificial matrix of semi-reality. It is semi-reality, because we recognize it as real, but it does not exist apart from the recognition and realization that we give it. Its entire presence consists of patterns of energy in the unified quantum field, vacuum fluctuation, that we modulate into the patterns that we recognize. Our entire empirical world exists because of the telepathic unity of human comprehension and visualization. It is real for us because we believe it is so. We recognize what we believe, and we do not recognize anything else. Prior to our incarnation, we agreed to limit ourselves in this way.

Quantum physics has assisted us in knowing that universal consciousness, the consciousness of the Creator of all that exists, is the source of everything. Nothing exists without consciousness as its essence, right down to the minutest of sub-atomic particles/waves.

As Self-realized Beings, we are constantly created in every moment within the universal consciousness of the Creator, which we participate in. Any of us can change the entire paradigm that we are living in at any moment with our focused intention, emotional power and imagination, provided that we've freed ourselves from all of our limiting beliefs.

No one can put us into any situation without our conscious or sub-conscious cooperation. We designed our beliefs to be almost unescapable from. We cannot even permit ourselves to question them, because we cannot allow ourselves to know who we are, in order to participate fully in the limited spectrum of vibrations of the human experience here.

When we decide that we intend to return to our full consciousness, we can use our creative abilities to do so. We must be serious about resolving all of our limiting beliefs about ourselves. We must keep examining ourselves for discordant energy within. And we must be motivated by the energy of our heart for the expansion of our awareness into the higher vibrations of unconditional love and joy. Here we can feel our participation in the life-stream of everything, and we can align with the increasing frequency of the resonant vibrations of the Earth.

Realizing Our True Genius

If we feel attracted to perceiving beyond our empirical sensations and apparently real world, we can look into the energy of our heart for guidance. It is here that our expanded and ascended Self connects with us. Higher guidance is ours to recognize. It is the frequency of unconditional love and universal consciousness.

In our empirical mind-set and belief systems, we cannot be aware of those vibrations. We are constrained by some level of fear, which requires our belief in our mortality. This is the deepest emotional knot that keeps us in the empirical spectrum. We know that belief in our personal mortality has been conclusively proved to be false by all those who have died and returned to their bodies and have reported about it. All retained their conscious personal being whether in or out of the body. From this we can conclude that our personhood is eternal, and we can change our belief about this, because we have now been logically convinced of the truth of our Being.

The proof of our Being is a quantum phenomenon, in which we are living in two different dimensions at the same time. We have our ego conscious person that lives in the spectrum of energies shared by humanity. We also have our expanded Being, who dwells in all spectra of vibrations with a perspective arising from unconditional love and universal consciousness.

Before incarnating on Earth, we experienced only high-vibration energy. We couldn't even imagine what it would feel like to live in low-vibrations. Now we know. We have deepened our understanding by being willing to experience the energies that we live within here. Once we realize that we want to expand our awareness, we are ready to open ourselves to the visions and emotions that feel really good. This is an individual soul journey.

We must conclusively confront our belief in our mortality in order to resolve all of our fears. As we do this, we can develop our sensitivity to feeling our intuitive knowing. This is the vibrational spectrum, mentally and emotionally, of peace, compassion and kindness. We are given prompts that we can recognize, along with feelings that we can be aware of. These may be subtle. They could be mathematical, symbolic, literal, musical or artistic and many other kinds of prompts. If we are open to receiving intuitive guidance, we will know it within. It arises out of universal consciousness and is true. This is our personal genius. We only need to recognize it.

How to Avoid Mental Crippling

The mind cannot know higher consciousness. The quantum realm is a mystery to the mind, although mathematicians have been able to conceptualize it; however, what they have found is a deeper mystery. We cannot comprehend infinity, even though we know that it's endless forever. Its recognition requires a state of serenity. This is the obstacle we face in trying to imagine operating in universal consciousness. It's actually something we naturally do, when we're unhindered by fear. We can pay attention to our feelings with the intent of achieving a state of serenity in complete openness of awareness. Here we can be aware of our intuitive guidance, which operates in universal consciousness in the quantum field. It is our natural knowing of everything and our guidance to create anything. In complete alignment with our intuitive knowing and being, we are unlimited in every way. In our true Self, we are each a pure self-aware personal presence of Being with unlimited abilities to express ourselves through creation.

The mind can be a useful navigator in our expressions and encounters, as we participate in human society and life. Since it cannot know higher guidance, we can intend to open ourselves to our intuition in every moment. The mind can help us to express what we know intuitively, and we can know everything we desire to know as deeply as we want to go. It is all within our consciousness. Everything we know is true comes through our intuition. It is how we know anything, and everything is within our knowing in clarity.

We can realize that everything we experience is a reflection of our personal consciousness. We are constantly creating the quality of every moment of our experience with our thoughts and emotions. We bring the energy of the quantum field into alignment with our own frequencies, and it materializes the energy of what we recognize, imagine and feel. As we desire greater participation in higher vibrations, we attract them into our experience. By consciously being in their spectrum of energy, we are aligning ourselves with a higher dimension of living and can be drawn toward our inner light.

We Are Creator Consciousness

Ancient bodies of teaching about life report that the cause of everyone and everything that exists originated with a Supreme Creator, a Self-Conscious Being. Out of this Conscious Being everything that has form arises. These teachings do not talk about time.

Quantum physics has now discovered the mechanics of how this can be. Physicists have found that everything that exists consists of energy patterns arranged within an unlimited number of frequencies and patterns. They call this realm the quantum field or vacuum fluctuation of all potentialities. The quantum field is the expression of a creating consciousness that is probably infinite. Everything arises from this consciousness and is created in the quantum field. This creating consciousness creates fractals of itself. Everything is a complete expression of its creator. Every sub-atomic particle/wave displays knowledge of its self-identity and universal consciousness outside of time and space. The quantum field is timeless and has no place. It is all things that can have an energy pattern or form.

In the quantum field all things concurrently exist now. There is no past or future. Every possibility is now. All creation is in the present moment, as is dissolution. What happens in our experience is what we choose to recognize. Nothing materializes for us, unless we recognize it consciously or subconsciously. Our recognition modulates the vibrations of the energy that we focus on, so that the energy patterns immediately spin on a subatomic level, creating the world we perceive as empirical. This world is held in recognition by the constant awareness of humanity.

The physical world is patterns of energy that we recognize. We can recognize what we experience as empirical. We can also recognize what we imagine we are experiencing. Because we are fractals of the Creator, we are constantly creating with our state of being, the vibratory frequencies of what we think about and feel. In our self-consciousness we are timeless, whether we realize we are in the body or out. We are the consciousness that creates everything.

Although we've limited our self-awareness to the spectrum of energies that we share with humanity, our true Being is the consciousness of the Creator. In our consciousness we can be as expansive as we want to go. We are unlimited and can express ourselves concurrently in many dimensions. Realizing our true abilities requires recognition. Our creative ability does not require effort, because it's part of our Being. The frequency of our energy signature attracts persons and situations that are compatible in vibrations. In this way we determine the quality of our lives.

Breaking Out of Our Limitations

In our essential Being we are creators and experiencers. With the life force that we constantly receive out of the Being of the Prime Creator of all that exists, we constantly create the empirical world that we inhabit. We do this by the focus of our attention, using the energy patterns of the quantum field that we recognize with our thoughts and feel with our emotions.

There is no solid reality in our empirical realm. Our physical senses stimulate our awareness in a way that makes everything seem solid, but in reality it's all energy vibrating in patterns of frequencies that our consciousness interprets as solid. By holding these energy patterns in our awareness, we create the situations that we experience. Our awareness is guided by our beliefs, especially our beliefs about who we are and what we're capable of. From infancy we are taught by our family, society, government and religion who we are to believe we are and what we are allowed to be. We do not have to accept any of this, unless we want to. None of it is part of our true essence of Being.

If we are spiritual adventurers and explorers, we seek release from our programmed beliefs and want to find out what our real potential is. Through deep meditation, breath work, inspiring music and art and aligning with the energies of nature, we can begin to free ourselves from our social programming. If we can be calm and serene within, we can feel and know the energies of our heart, our innate intuitive knowing of our personal truth. We can feel the love and joy flowing into us through the life force streaming out of the conscious Being of our Creator. We can begin to live beyond our limiting beliefs.

In our intuitive knowing we can open our awareness to our eternal Being, unlimited by time and space and empirical stimulation. In our consciousness we can travel out of our physical body awareness into the cosmos of unlimited Being. We can become aware of our sovereignty and absolute freedom of Being. We can create anything we can imagine and feel. This is actually what we do all the time without realizing it. Every night we travel out of the body. We are constantly creating our experiences in the world that we recognize, just by the nature of our Being. We are free to change our situation in life at any time through our imagination and emotions.

In our essence we participate in the universal consciousness of our Creator, and we are capable of realizing this, once we release our limiting beliefs about ourselves and learn to recognize the vibrations that we love the most.

Attaining an Enlightened State of Being

We're learning to ride the waves of life on this planet with the awareness of playing a game, not having our mortality at risk. We're gaining the ability to rise into our eternal Self-awareness. We can look at life as a metaphor for our understanding. Everything that happens in our world is significant on a symbolic level. As directed by our One Creator consciousness, our experiences are designed to elicit in us realizations of anomalous and discordant energy patterns that we hold. These manifest in our psyche as limiting beliefs that stream from the belief in our mortality. As we recognize everything tinged with fear, we can resolve these energies by pouring our high-frequency vibrations of love and compassion into every challenging situation. As we are not in alignment with fear-based energy, it disappears from our experience, leaving space for high-frequency energy from our life force, which enables more expansion as we open to it.

We are learning to conduct ourselves in alignment with a frequency spectrum closer to the consciousness of the Creator. The path toward the light in our consciousness is being enhanced as our life force builds along with the expansion of our awareness. We can become aware of the photons being released in our heart and radiating around us, attracting resonant energy patterns for us to experience. We are Beings of light, we emit light. It's from the life force that flows into us constantly from the universal consciousness of the Creator and modulated in our personal consciousness for our experience. We can recognize ourselves as personal Creator Consciousness. With our continuous thoughts and emotions, we are constantly creating the qualities of our experiences in this world by the frequency of our energy signature. We control our energetic resonance with our attention, our imagination and emotions.

By being aware of the quality of energy in our presence, we can recognize what it is and change it in our imagination and feelings, so that we recognize it as divine energy patterns aligning with our high-frequency resonance. We can feel it as angelic energy. As we learn to intentionally change the vibrations of a situation to align with higher frequencies, we enter the mystery of the quantum field. Here what we recognize, and what we believe to be, manifests as physical experience. It doesn't matter what it is, or how unlikely it might be. When we become adept at this (usually after a long and intense period of practice), we appear in the empirical world as if we are not subject to its boundaries and apparent substance. We can make our physical body appear and disappear. We are unlimited in universal consciousness.

Chapter 6. Universal Consciousness and the Quantum Field

Enlightened Relationships

Most humans live with much personal drama in relating with others. We have strong likes and dislikes, fears and desires. We see ourselves as separate from others by religion, culture, politics, race and even chemical injections. Many believe that we are inherently sinful and untrustworthy. They believe everyone is like this, and that we need a savior to let us keep on living much as we do now, but better, and not have to deal with the consequences of our thoughts, feelings words and actions. That is an impossibility, because everything is energy. In our true Selves, we are Beings of radiant light and Joy and unconditional love. We are creating our human lives in every respect through the frequency patterns of our conscious and subconscious imagination and emotions in every moment. It is our predominant frequency spectrum that attracts compatible energy patterns that resonate with ours, within the boundaries of our beliefs.

When we vibrate at low-frequency, we cannot tolerate high-frequency encounters, because our vibrations become anomalous, and we become unstable, with great physical discomfort. Only by raising our vibratory levels can we rise into more expanded awareness. This is an intentional process. We must come to know love, joy, gratitude and compassion. We must feel at One with Gaia, the Spirit of the Earth, and vibrate in resonance with her.

As we progress along the path to awakening, we resolve all of our limiting beliefs about ourselves, so that we can be clear. In our true Self, we are eternal Self-Realized Beings with personal, present awareness and awesome abilities. We have everything we could ever need or want within the reach of our creative abilities. We are energy modulators with our imagination and emotions.

Yet we are all human, all conscious Beings arising from the One Creative universal consciousness that some of us call God or Allah. Without personal limitations, we can be free beyond time and space. We can consciously participate in universal consciousness. We can experience feeling that we are the same divine Being as everyone and everything. There is One consciousness that constantly emits the creative life force that enlivens everyone and expresses everything as electromagnetic energy patterns, which we consciously interact with. Our intuition relays to us the quality of vibrations that we want to align with in every moment. When we are fully aware of our intuitive knowing, we can create beautiful lives aligned with the high vibrations of an ascended dimension. Here every relationship is loving and expansive.

All of the low, fear-based, vibrations exist in a lower-frequency dimension and are not present in the higher dimensions. With practice and clear intent, we can live in the higher dimension now, even occasionally, as we are able. This is where we meet one another in our true Selves.

Participating in Universal Consciousness

In opening ourselves to more expanded awareness, we can find many aspects of nature to help guide us in our inner journey. Our planet has an amazing variety of environments and beings that inhabit them, and all of them offer an awareness that we can align with for greater understanding of life. We can begin with Gaia herself.

Our planet has adjusted her resonant frequency to that of humanity for thousands of years and was nearly destroyed by our low vibrations. Now Gaia has begun to change dramatically, as she aligns with the higher-frequency photon cloud that we are enveloped in and the increasingly powerful gamma ray emissions from our Sun and our cosmic environment. The Shumann Resonance graph of Earth energies shows the Earth's vibrations now occurring on four higher octaves of frequencies than has historically been known. In order to continue to live here, humanity's resonance must also rise, or we will become energetically unstable in our life forms.

By spending time in nature and just being aware of as much as we can, we can immerse ourselves in the resonant energy patterns of everything around us. We can feel the soil, the sand, the rocks and perhaps even witness lava flows. We can listen to the sounds of the creatures around us and begin to feel the energies they radiate. We can smell everything around us and absorb the energy of the Sun, the wind and the rain. We can go deeper into our inner knowing.

On our deepest inner knowing, we can identify with everything we focus on and can align our awareness with the essence of Being of all aspects of nature. By going deep within, we can feel the radiant energy of an animal, a plant or a rock. We can know how it feels and what it is aware of, to the extent that it has those capabilities. We can know its essential Being, just as we can know our own. There is no difference in our self-consciousness of who and what we are. There are differences in our capabilities and abilities to express ourselves, but in our essential Being, we are all the same consciousness and can share awareness of everything that we are.

In our awareness, we can enter the Spirit of the Earth or any conscious entity. We can be the tree, the bird, the dog, even the amoeba. It all depends on how deep we allow ourselves to go into our consciousness awareness. In our essence we are unlimited, once we clear all of our self-imposed boundaries to our awareness, which we share with humanity.

At the heart of everything we can discover the flow of life force from the consciousness of the Creator and the unconditional love that connects every expression of consciousness.

Remembering Our Personal Truth

On the path to remembering the eternal limitless Being of our true Selves, we have the ability to transform every obstacle, because we continue to create the obstacles. We can change our perspective through aligning with the energy of our heart and our deep intuition. Ultimately we can be fully aware of being in universal consciousness with its unconditionally loving life force flowing into all conscious beings. This perspective allows us to be our creative Selves, expressing ourselves out of compassion and deepest love. By being intentionally present in this high vibrational state of Being, we remember that we know everything. Every Being within the spectrum of our personal energy signature is known to us. We all are constantly being created out of the same universally conscious Being. We are all creating experiences for ourselves and the Creator, constantly expanding consciousness.

Since we consciously and subconsciously choose our focus in every moment, we have the ability to elevate our conscious frequencies by clearing our beliefs about ourselves. These constructs of our ego are no longer needed, because they're all based in fear, ultimately fear of termination of consciousness. We know from quantum physics that there is only One consciousness. It is the eternal presence of self-awareness, always now, beyond time and space. This is the Source of our Being. It cannot be terminated, because it is always present, encompassing and enveloping everything. It is the Source of all of the electromagnetic wave/particle patterns in the limitless quantum field of all potentialities.

Fear of termination is based on a false premise. Since termination is impossible, fearing it means that we recognize it in our imagination and create it for ourselves. It becomes our belief and keeps us in the spectrum of fear. Our life force continues to create our experiences, based on our beliefs. What we believe is always true for us, because we've created the vibratory patterns.

Recognizing that our conscious awareness must be eternal, gives us the freedom to focus on the spectrum of vibrations based on love and compassion. We can be aware of the feelings emanating from our heart and the guidance of our deep intuition. This is the natural flow of our lives that we can align with.

Knowing Creator Consciousness

In the process of getting to know who we really are as conscious, living persons, we can open our awareness beyond the physical realm to our consciousness itself. We can observe ourselves from another dimension with greater consciousness. In our true Being, we can express ourselves in every dimension timelessly.

In our human being consciousness, we live within the limitations of the empirical spectrum of frequencies. We have compartmentalized our consciousness to be able to participate in the human experience. When we are ready, we can withdraw our attention from this world and focus on the energy patterns that enhance our lives. These are the vibrations that we know and feel in the heart of our Being. They are all based in the spectrum of love and joy. This is our natural state of Being. it is the essence of the quantum field. When our thoughts and emotions vibrate in alignment with these frequencies, we know how this feels in our heart. This is the energy that we can live in as long as we choose to. We modulate this energy with our imagination and feelings, which create our personal energy signatures and the resulting resonant experiences.

Living from the heart of our Being draws us into our realization of the presence of the Creator, our greatest Being. In the heart of our Being we receive our eternally present personal life force. We can recognize our timeless presence of Being as a fractal of the One Creator. We are the Creator. We are the presence of the Creator, expressed as our persons, and enveloping in our consciousness all that exists.

By aligning our attention with love vibrations, we become more radiant in resonance with our natural spectrum of emotions. We can face any energy patterns that could cause instability in our vibrations, if we can maintain a perspective of love and compassion. We can be aware of the misalignment without ego involvement. With this kind of awareness, we can transform all lower-vibration energy patterns into the spectrum of love and joy. We are the modulators of the energy.

Our Backwards Process of Creation

As we experience the circumstances of our lives, we have learned to believe that things happen to us because of forces outside of our own conscious being. We have believed that our condition in life and the way others treat us is imposed on us from outside ourselves, that we are subject to forces beyond our personal control and are victims of circumstance. None of this is true.

We are sovereign Beings participating in the universal consciousness of our Creator Being and participating actively in every moment in the creation of our lives in this dimension of experience. How can we know this? We have learned from quantum physics that there are mysterious forces that we cannot understand from our empirical perspective. Physicists have identified the universal consciousness that constantly creates everything in expression of intelligent patterns of energy in a plasma realm that is understood as a unified quantum field of all potentialities. This is our cosmic environment, and it envelopes us and provides all the energetic patterns that we recognize as our experiences. Everything is conscious, from the minutest subatomic waves/particles to galaxies and universes. It all interacts with us according to the energetic patterns that we recognize and feel in our thoughts and emotions. Our conscious recognition creates our experiences.

As we observe and recognize energetic patterns in the quantum field, they become experiential in our physical bodies or in our imagination, but we have our limitations. We can recognize only what we believe is real for us personally. By believing and feeling that we are victims of circumstances, we become victims of everything from political mandates to poverty and suffering. By believing and knowing that we live in freedom, love and abundance, we create the circumstances that we can recognize as providing this reality in our experience. Once we recognize the energetic patterns that we want to experience, we are prompted in our intuitive knowing to take action that will be in alignment with the vibratory frequencies of our visions and feelings. If we do not follow through with our knowing, emotional creativity and inner guidance for our physical activity, our visions cannot become our experiences. If we falter in our creative process, we experience whatever we believed and realized consciously.

Our current situation is always a result of the energetic patterns that we put into the quantum field through the vibrations of our energy signature. These vibrations magnetically attract energetic patterns that align with us in our thoughts and emotions. What we believe we are experiencing becomes for us what we experience.

7.

Aligning with Higher Consciousness

Love as All-Encompassing Reality

What happens when we realize that love is all around in everyone and everything? We're all playing free-will-chosen roles in a world-wide drama for our experiences and the satisfaction of our curiosity without Self-realization.

We have learned that our reality, our personal experiences, are a result of what we recognize. Everything exists as energy until we recognize it as a projection of our own conscious realization. We are the creators of our reality and everything about it—the entire empirical matrix. Once we realize this and submerge ourselves in this perspective, we free ourselves from all of the boundaries and limitations we have experienced. It increases our joy naturally, knowing that we live in a vibrant and fulfilling world. For most of us, this doesn't just happen suddenly. It takes intention and practice and constant awareness of our emotional state.

When we recognize that we can be living a truly wonderful life, we can begin the process of transforming ourselves. Our emotions are very important in being able to align with the high-frequency energy of true love. These are the feelings that are natural in this realm. They are the gift of the Divine Feminine, and they require intentional focus for those of us who don't automatically fall into divine love. This is the feeling to maintain as much as possible, until it becomes natural in our sense of being.

The deeper we go into love we increasingly recognize our greater limitless consciousness, flowing constantly with unconditional love into our Being out of the quantum field in the divine creative consciousness. To actually realize this is an ecstatic experience. This is the experience that we will always know, because once we know this, we cannot unknow it. From this time onward, we will be attracted to ecstatic love, and we will feel our life force flowing into our being from the quantum field of our Source consciousness. We will be aware of our natural state of creative being and will focus on high-frequency energy patterns and use our imagination to navigate the quality of the experiences we're interested in.

Eventually we feel from within that everything is the same Being, individualized for personal experiences, and all arising from the same encompassing Self-realized consciousness. From here we can create the most wonderful lives for ourselves and everyone.

Living in Transcendent Vibrations

What is our greatest joy? How do we attain it? The design by which the cosmos exists is for all entities to operate precisely in synchronicity. Humans have the option to violate this, because we live in an imaginary realm. We cannot fatally damage the structure and operation of nature, unless we do so on atomic and cellular levels and engage in mass destruction. This kind of super-low-frequency energy is being recognized now, and its propagators are trying one mass extermination attempt after another, in order to hold off their extinction, which is imminent. Their boundaries in consciousness are closing in around them with their energy signatures breaking up. One by one they are becoming unstable in their being, and are disappearing. The low frequency energy is dropping away from humanity, and many are now realizing that we're all part of the *Trueman* movie.

We can remove our boundaries and transcend our entire dimension. We do this with our consciousness. We become aware of our own presence of being. Each of us is all there is. We encompass the cosmos of universes in our consciousness to the extent that we are willing to go. Our experience as humans is a small part of our eternal Being. We can now be ready to identify with our true divinity, the energy of life constantly pouring into us within the quantum field of universal consciousness and unconditional love. By resonating with this energy we penetrate the level of consciousness of the higher dimensions. Our true Self has no limits, except for our self-conscious presence. We are completely fulfilled in every way. We are filled with gratitude and deepest love and are in the presence of our soul mates. This is the greatest joy, and we attain it by realizing its reality for us.

The dimensions all exist concurrently within the quantum field. We each have portions of ourselves in other dimensions, all contributing experiences to our essential Being. We have gone on adventures to gain experience and wisdom, as well as enjoyment. One of those adventures is our present life on Earth, living in this artificial energy pattern of the human energy signature that we constantly create by recognizing it and feeling it.

If we are to raise the energy signature of humanity into a higher dimension, we can do it individually through our recognition and realization and feeling higher vibrations. We see this happening on the Shumann Resonance graph of the vibrations of the Earth. We're aligning ourselves with the natural energy of the higher dimension. We magnetically draw that energy into our experience and create more light for humanity.

The Limits of Our Intelligence

With the recognition of quantum physics that universal consciousness is the basis of everything that exists, and that everything participates in the one consciousness within the individual identity of all entities, we humans have our own unique identity and consciousness. No limits are imposed upon us. To the extent that we can open ourselves to the unlimited stream of conscious life force that enlivens us, we also open our emotional and mental intelligence and awareness beyond our accepted limitations.

Scientists have found that what we consider solid reality is really mostly space. Quantum physicists have found that it's plasma, part of the unified quantum field. The atoms and molecules are also mostly plasma. The sub-atomic particles occupy very little of the space. Yet we perceive our world as solid. Its design is a creation of the human imagination, held in form by all of us who recognize it. It is our consciousness that creates the reality for us.

Other realities can exist simultaneously in our awareness. They become real for us when we recognize them. They exist in their own spectrum of energy, just as our normal human world does. As we personally raise our innate frequency, we experience higher dimensions that we resonate with.

As humans on this planet, we have created deep-seated beliefs to limit our awareness, so that we could experience a low-vibration world as real. Once we feel that we've had enough, we have reason to awaken to a more expansive reality. The path to that awareness is different for all of us. If we look at it from an energetic perspective, we can imagine how it can work for us to get to a life closer to unconditional love and joy.

We are approaching our awareness of our connection with the constant source of our life. We are enveloped within the unified quantum field. Our energy signature, the vibration of our state of being, attracts encounters with other beings within our personal spectrum of vibrations. Everyone here is present because we all vibrate within the spectrum of humanity's frequencies.

We are multidimensional beings. We can project our conscious awareness wherever we want by changing the vibratory frequency of our energy signature. To make the leap in frequency that is necessary to resonate in a higher dimension, we can go back and forth between dimensions until we get ourselves settled in the higher frequencies. At that point, we'll be walking on water and through walls. Energetically we just need to connect with the water molecules and not the space between them, when walking on water. When walking through walls, we need to walk through the space and avoid the atoms and molecules. This is a matter of intention and absolute belief and knowing our true Being, expressing itself as us.

Our true Being is an expression of the infinite Creator, who expresses itself as the unified plasma quantum field of all potentialities, out of which we arise. The quantum field has universal consciousness and infinite creativity, modified by our conscious presence. Once we drop our false beliefs about our limitations, we can recognize our true unlimited, eternal and unconditionally loving Self.

Expanding Our Understanding

Once we are predominantly living in high vibrations, we realize that we've been involved in an alternate reality that is very convincing. It's a trick we've played on our own consciousness to experience being limited to this alternate reality of the human spectrum of vibrations.

We've been involved in a limited, low-vibration alternate reality, a game in our consciousness to pretend that we're limited and mortal. The game is designed to be intensely involving and fear-inducing. From this limited perspective, we can disconnect by staying mentally and emotionally attentive in high-vibrations. As our conscious awareness becomes sharper and more expansive, we understand the situation on this planet, and we can have compassionate wisdom in all situations.

We can raise our vibrations to be able to live in a high-frequency spectrum of experiences. We do this by living in gratitude in all situations. We can be in gratitude, while feeling joyful and compassionate. This is a natural perspective of high-frequency emotions while knowing that we are unlimited in our conscious Being.

We have compelling experience in the presence of our Being. We become aware that we are personalized fractals of the One consciousness that creates everything always. We are timeless and without location. We are pure conscious presence, with many unlimited abilities, the same as the Creator. Each of us is the creator. We can play with all the energies around us infinitely. We've been doing this in the human experience constantly. We've just kept our consciousness within in the human spectrum of frequencies.

Moving our consciousness beyond the body is easy for us. We do it all the time in day dreaming. We can intentionally day dream experiences of beauty and joy and everything that resonates with those vibrations. We can transform our perspective from wherever it is now to recognizing a higher, more expansive reality interpenetrating our human experience in a multidimensional quantum field.

Chapter 7. Aligning with Higher Consciousness

Building Escape Velocity

We can realize deeper aspects of each other, and be more conscious of feeling the presence of each other, whether present physically or not. We can feel the radiance even at a great distance. We can feel the energy signatures of each other. We feel vibrational patterns. We can move through the anomalies of personalities to the essential Being of each of us. We can resolve all of the low-vibration energy patterns that arise by staying in a high-frequency spectrum, the realm of compassion, love, joy and wisdom.

We are not bound by the Matrix of humanity's energy spectrum, unless we choose to be. There are alternate realities all around us in different energy spectra that we can become aware of by aligning our imagination and emotions to a spectrum of frequencies that we want to experience. We express ourselves as energy beings that appear in a spectrum of vibrations that stimulate our senses in a way that our consciousness interprets as physical or empirical.

The world of human experience consists of patterns of electromagnetic energy waves that we recognize as the real world. It's no more real than any other world that we can imagine, except that all of us are focused on this spectrum of energy. We exist in a unified quantum field of all potentialities. Any scenario is possible. We are the creators through our ability to modulate the wave patterns in the quantum field with our thoughts and emotions. We determine the frequency spectrum of energy that we live in, and the quantum field gives us infinite potential.

We can navigate successfully through life by paying close attention to our emotions. Whenever fear arises, we can change perspective to resonate with high-vibration energy patterns that enable us to resolve our experience.

We can expand our conscious awareness beyond the spectrum of the human Matrix of energy, while at the same time raising the energy signature of humanity by living in high vibrations of thoughts and emotions.

Unleashed Beauty and Joy

As embodied beings on Earth, we have agreed to subject ourselves to all of the limitations to our conscious awareness that are within the frequency range of the energy spectrum of humanity. Those limitations are now to be transcended by all who have had enough of the spectrum of energy that humanity has operated within. Humans are graduating to a higher-frequency spectrum of energy and life.

We are returning to a state of recognition of our essential Being. We can come to a state of peacefulness and emotional neutrality with the help of deep, rhythmic breathing while at the same time feel ourselves filling with greater awareness of our Selves. This feels like an expansion into deepest love and gratitude. It feels like compassion for our ego self for all of the struggles we have gone through as limited-conscious beings. By resolving all of our beliefs about ourselves, as they come into our awareness through challenges to them, we can understand ourselves through compassion and wisdom. There is no fault, judgment or blame. It has all just been experience for us to know what it's like to live in the lower vibrations of life as humans.

Now we have the opportunity to live in a high-vibration human experience. We get to this level of vibrations by being in high-vibration emotions and imaginings. It helps to be in nature, especially in spectacular places, calm places, high-energy places, whatever is attractive to us for centering and balancing ourselves in alignment with the energies of the essence of the Earth. We all have our own ways of finding this inner space of just being present in our awareness. Here we have no boundaries or limitations. As we open ourselves to infinity, our consciousness can expand as far as we want to go.

Together with the Spirit of the Earth, we are creating a fully-regenerated most beautiful planet of living energies in a wonderful dimension. This is where we can live in peace and kindness, sharing the great consciousness that creates the unified quantum field and everything that exists. The Creator extends Him/Her/It Self as us. Each of us is a fractal expression of unlimited conscious awareness in a state of unconditional love and joy.

Transforming Humanity

We have reached the very end of the End Times of the low-vibration patterns of humanity. The natural energy patterns of Earth and our galaxy are rising in frequency, as exemplified by the Shumann Resonance graph. All of humanity's energy blocks and negative emotional knots must come to light and be resolved through compassion and wisdom. The low-vibration energy must come into alignment with higher vibratory patterns or become unstable in its frequencies and wavelengths and dissolve. The parasites must transform or leave our planet.

We are the creators of high-frequency personal vibrations. Through our high-vibration energy signatures, together with the rising vibratory rate of the Earth and our nearby cosmos, we are raising the vibratory rate of humanity. We do not give any credence or life force to the current dramas of humanity. They are only manipulated energy patterns that do not affect us, unless we allow it. Historically almost all humans have accepted it, but now conditions have become so insane, that people are realizing that we really have been slaves. Innately, however, we know that we are sovereign beings and are beginning to awaken to our true Selves.

Any energy patterns that we face, we can transform through our intentions, imagination and emotions, if we are absolutely present and deeply know our true Being. This usually doesn't happen all at once, but we grow into our Selves with intentional practice. This process happens for us, when we live in the energy spectrum of compassion, joy and love. They attract experiences that are compatible with their energy patterns.

Living a High-Vibrational Life

Currently many of us are working through low-vibration emotional attachments that have kept us enslaved to low-vibrational lives of restricted living conditions. There are many ways of resolving these situations. One is to strive passionately for awareness of the highest dimensions that we can imagine. Another is to take a small step toward enlightenment and continue this process to the goal of complete inner clarity. The important perspective is to search for higher and higher emotional vibrations and inner experiences.

By focusing on someone and feeling and looking for the inner light and love, indistinct though it may be, we can telepathically and empathically direct love and joy, regardless of what that ego and physical personality may be. We can approach all low-frequency situations and emotions with our high-frequency energy signatures. If we maintain our high-frequency feelings, the lower-vibration energy will have to come into alignment with us or become unstable and dissolve, leaving any beings who depend upon our life force to expire from our experience. They terminate themselves by intentionally cutting themselves off from the life stream of the Creator flowing through the unified field of quantum fluctuation and all potentialities.

We are not required to contain our self-awareness to within the existing human energetic frequency spectrum. We have complete control over the focus of our attention and our emotional expressions. These are our tools of conscious expansion. We can also feel that the natural flow of higher-frequency energy of our planet, which all of us must stay in alignment with.

We can realize the full force of the creative power of our divine Selves by recognizing our true innermost Being and feeling the unconditional love that flows to us constantly. We can know our Selves through our recognition and emotional state of Being. This is who we are in our essential Being, apart from all of the false beliefs about ourselves. We are eternal sovereign Beings with the ability to modulate vibrations of energy patterns in creative ways.

Transformation of History

As humans on this planet at this time in history, we are being drawn into the compression of time and rising resonant frequencies. Our lives and everything we have known are being upgraded to align with a more refined sense of being. This is true for us individually and for all of humanity, as well as for the Earth. Our skies are being cleansed and are returning to their natural deep blue with supportive weather patterns. The oceans are being purified of radiation. Forests are returning. Extinct species are being recognized again. All of our social and financial systems are transforming. We're in the beginning stages of this transformation, and the changes will accelerate as we move forward into a time of greatness and celebration.

The Earth is ascending into a higher consciousness, and our galaxy is sending us massive clouds of gamma ray conscious photons to transform our DNA and elevate out consciousness. As free-will beings, we have the choice of participating in this energy of renewal and spiritual transformation on a regenerated planet or remaining in the spectrum of vibrations that has pervaded our planet for eons and continuing without interruption on another planet with the energetic patterns that the Earth has experienced for thousands of years.

Our transformation is dependent upon our opening to the truth of our Being. In our own consciousness we are being attracted to the higher vibrations of unconditional love, abundance and joy in our essential Being. This is the quality of the natural energy patterns experienced in the quantum field and expressed as our life force and expanded consciousness. We have closed ourselves off from this realization in order to live within the energy spectrum of humanity as full participants. We have locked ourselves into a low-vibration world of experience. Once we recognize an inner prompting to want a higher-vibration life of more love, we can begin to feel for higher-vibration experiences in our imagination.

When we have learned to live mostly in the realm of compassion, love and joy, we attract mostly experiences of the same quality of vibration. The rising resonance of the Earth and our cosmic environment are drawing us into lives of higher vibrations, making our path to our Self-realization more attainable. At the same time, this natural energy is destabilizing the low-frequency realm that has been humanity's spectrum of energy. Those inhabiting the lowest-vibration lives and have lived out of sight are having to gaze into the light. They will either come into alignment with love and gratitude, or they will decompose. As the old world passes away, the new world is getting brighter and more attractive.

The Enlightened Ego

Upon incarnating into the energy spectrum of humanity, we create an ego consciousness that enables us to participate fully in the human experience. Prior to birth, we decide what that experience will be. We give ourselves all of the conditions and abilities that will best enable us to fulfill our intentions. We create our own destiny, but we have the free will to change the quality of our lives at any time. We do not predetermine how we will react to any situation, or what we may create with our thoughts and emotions.

We designed our ego to live within the energetic boundaries of humanity and to do its best in keeping us alive in the body and prospering whenever possible. We wanted to have every possible experience of fear, however subtle or powerfully overt. The ego has brought us to the present, still in the body and aware. It has had to perform without higher guidance. We can be deeply compassionate for its performance. It has given us the experiences we wanted, so that we can have a much deeper appreciation for the qualities of low-frequency living. Now we can have much deeper compassion and understanding.

We can love our ego as our own creation, providing us with the opportunity to penetrate deeper into the infinity of consciousness. We can assure our ego that it may now relax and enjoy a larger and more loving state of awareness of Being. We can attune to the energy of our life source through the intuition of our heart. This is the realm of the vibrations of joy, peace, abundance and freedom. To be in this vibratory pattern, we must be present and aware of subtle promptings that we know.

Our feelings reveal the vibratory frequencies of everything in our experience. The mind cannot read emotional energy. Our emotions are aware of their own energy patterns and are able to create emotional energy patterns. By increasing our brightness in love and joy, we are able to believe in the new reality. Our ego now has an easy life and is enjoying being cared-for.

Manifesting Our Visions of the New World

Since everything is electromagnetic energy patterns, any changes that we create must attract energy patterns in resonance with our creative vibratory frequencies. By imagining and feeling ourselves participating in high-frequency scenarios, we are creating experiences in resonance with these energetic frequencies. We can do this with all aspects of life, spanning as far as we desire. As each of us practices this form of creation in love and compassion, we can intentionally enter the presence of our true Being.

While we are moving into living in very high-frequency unconditional love that constantly flows through our heart on every vibratory level of our Being, we are also attracting others who are in resonance with our energy signature, strengthening the power of our radiance into the quantum field for manifestation in our experience and radiating into the energy spectrum of humanity, motivating others to awaken to their real Selves and recognizing the contrived structure of the Matrix we've been confining our awareness to. Our expansion of consciousness leads to others' desire to participate in this process. We already have the critical number of awakened souls to be able to transform humanity. This is what we are doing intentionally, as we create the world of high-vibratory joy, abundance and freedom in all ways.

The low vibrations of fear cannot exist in a high-vibration environment. Low vibrational anergy must come into resonance with the higher frequency emotions or become unstable and dissolve into random energy patterns, waiting to be modulated by a creative being. This is the process that we are introducing to humanity by expanding our realization of who we are. As more of us come into alignment with high-vibration realization and feeling, we are expanding the consciousness of humanity as well. We interact with one another from a perspective of compassion and wisdom, creating high-frequency interactions.

By taking time to meditate on conscious expansion, ultimately to encompass the cosmos, we begin to recognize the unlimited presence and feeling of our own Being. Now we can encounter life from this perspective and expect miracles.

Living in Alignment with Our Truth

So much of our experience among humanity is fake, including

our beliefs of who we are. There are good reasons for this, and we are ultimately responsible for all of it. What matters now is realizing the essence of our Being and the energy patterns that we create for ourselves. We are in the process of creating an entire new world from nothing. The current Matrix of our experience is dissolving. The rule of military force is ending. The banking oligarchy is terminating. Poverty is about to disappear. We are becoming our own commanders of our lives.

New assemblies are forming all over the world, soon to replace the destructive governments that have suppressed us for eons. New medical technologies are being introduced that can heal us of everything. We have a brilliant future, supported by the rising natural energy vibrations of the Earth and our galactic environment. We're consciously transforming the realm of low-vibrational existence into an all-encompassing experience of peace and joy.

As we inspire one another with creations of beauty and majesty in all areas of life, we are coming into resonance with the high-frequency vibrations of our Creator. We only have to follow what feels best to us. We all know what inspires us in love and happiness. These are the energies that we all naturally want to live in always. We have the free will to choose to live this way. By choosing to interact with one another in love and inspiring ways, we are following our intuitive guidance into the new world of joy and abundance. We can realize the truth of who we are, and what it feels like, once we have trained ourselves to believe that we truly do share the fullness of our Creator in our expanded consciousness. This is what Jesus meant when he said that we can do everything that he did and more. All who choose to do so are opening into divine awareness of our illuminating life.

Our Situational Awareness

What keeps us from recognizing our infallible inner guidance in every moment? The primary factor is the beliefs that we have been taught and have accepted without higher guidance. These keep us from being open to anything that is incompatible with them. There would be lack of resonance and resulting instability in the energy patterns that we live in. Every time we're involved in a traumatic situation, we experience instability and stress. Many situations involve constant stress and struggle for physical survival. These are all situations that are incompatible with the flow of our natural conscious life force, but we experience them with fear, because we are unaware of our intuitive guidance.

When we believe that we are separate from the Source of our life, we become untrustworthy and vulnerable. This perspective attracts experiences that elicit further compromise of our freedom of Being. We become enslaved to the operations of our ego consciousness, beyond which we do not allow ourselves to become aware. As long as we believe ourselves to be mortal, we cannot be aware of our eternal Being.

We believe that we are not divine Beings, and so we are not. We become sinners in our self-realization, and we look to a separate God to redeem us, to offer us atonement (at-one-ment). In our reality, we are constantly created as One with the prime Creator, fractals of divine Being. The entire universe is fractalized essence of the Creator in the form of Planck particles/waves.

Our guidance comes from our intuitive knowing. It is part of our natural life force, which flows through our heart and energizes us. We can align ourselves with this spectrum of vibrations by observing and feeling the rising energetic vibratory patterns of the Earth. These are subtle, and our emotional awareness, which is part of our intuition, feels them. Our intuition uses these feelings to communicate with us and will fill us with elevating feelings when we're using our abilities to create loving and joyful high-vibration energy patterns that resonate with our natural conscious life force. From our natural high-vibration perspective, we radiate this energy into the quantum field and into the energy spectrum of humanity in our state of being and in our actions. This creates a process of resolution of the energies around us and transforms our experiences.

Realizing our Essence and Abilities

We were created to provide inspiring experiences for our Creator. As fractals of the One Conscious Being Who is everyone and everything, we are the creators of all of our experiences through our ability to recognize new forms and vibratory patterns out of the electromagnetic field of energy that envelops us. We have the free choice to create whatever frequencies and combinations of wave patterns that we desire through our focus of attention and emotions. Our beliefs enable us to recognize what we focus upon, which then appears in our experience.

By resolving our beliefs about ourselves, we expand our awareness. We have been subjected to cultural and religious training that we have believed, because we want to participate fully in human society. We generally believe that we are subject to what appears to happen to us without any intent on our part, especially events that are catastrophic and life-threatening.

Nothing happens by chance. Every situation that we face is a result of energy patterns that are attracted to our personal energy signature, which we are constantly shaping with the thoughts and emotions that we focus on and express. If we are fearful, we attract fearsome situations. If we are joyous and kind, we attract compatible situations. Everything consists of interacting energy patterns that we consciously and unconsciously create and interact with in our state of being.

We are Beings of light. We are photon emitters. Those of us who can see beyond our shared visible spectrum can recognize our aura of light. It is the light that arises for us with our conscious life force out of the conscious Being of the Prime Creator, the universal consciousness that we are enveloped within. Our conscious awareness can encompass more than the universe. To be in this presence of Being, we only need to follow the guidance of our heart-felt intuition without any limiting boundaries of beliefs that we have accepted about ourselves. We must release the belief in personal mortality, which has been conclusively proved to be false by all those who have died and come back, and who have reported on their experiences beyond body-consciousness. All have disclosed that their self-aware personhood was unaffected. They all experienced being the same persons, whether in or out of the body. We are eternal Beings of universal consciousness, and with unlimited ability to resonate patterns of energy that become the qualities of our experiences in the body.

We can create everything we can imagine and empower with our emotions. Our experiences in the empirical world are all reflections of our own consciousness, and all of us are the same Being, all arising out of the consciousness of the Creator, and all being Fractals of the Creator, participating in the entire consciousness of the Creator. This is true for all conscious beings. The human uniqueness is the ability to change the energy that we encounter. We are not just Beings, we are the Creators in every respect.

Our Greater Being

We live in a realm of duality in vibrations. There is a division at zero-point in vibration and emotion, between the realm of fear and the realm of love, between low-frequency vibrations and high-frequency vibrations. These vibrations correspond with the emotions that they stimulate in us. We have the free choice to focus on and feel any of them.

Our low-vibratory spectrum of life with humanity has given us a depth of experience that would not have been possible, if we lived only high-frequency lives. Our compassion and understanding are much deeper now. We are destined to use these qualities in our new creations, in choosing the vibrations we

Chapter 7. Aligning with Higher Consciousness

want to pay attention to and experience. There is no requirement for us to fix our vibrational focus to our current situation. We can move freely between the two realms of fear and love.

Our challenge has been to be able to change our perspective to high vibrations while being stuck in a low-vibrational environment. We can train ourselves to do this. One way is to imagine ourselves to be our divine Self in our current situation. We can open our awareness to greatness in every being. All have an inner light that emits photons to display a subtle aura of light in a higher vibratory range, that some of us can see, and that technology can detect. With the inner light comes conscious life force, which fills the essence of all beings. We can realize that we arise out of the same Being with everyone. No one is separate from each of us. We are all expressions of the One Creator Consciousness. Each of us is a personal presence of Being, constantly arising from the Creator.

We are quantum expressions of Being. We can be in more than one place at the same time, because in the quantum realm, there is no time. We are pure Being and can express ourselves in many timelines concurrently. Human life is one of them, and we exist in many others right now. Our consciousness is vast. To realize it, we must turn inward to our deepest knowing, where we can feel our eternal essence. Here is unconditional love and joy flowing through our heart. Once we experience this, we no longer need to believe in our mortality, and we are ready to release our attachment to the realm of low vibrations.

Realizing Our Potential

Our planet is rising in resonant frequency and is releasing much negative, low-vibration energy through extreme weather and earth changes. This energy is flowing through all of humanity. Low-vibration alignment is becoming very difficult. The corrupt ruling elite are leaving our awareness, and their unseen leaders are dissolving. The resonant frequency of humanity is rising with the Earth. The days of love and goodness are immediately ahead of us, but the path to this level of vibrations will be volatile. There is much discordant and anomalous energy to realign with truth.

The present energy of Gaia's current path is guiding us into higher vibration living, more in tune with the energy of our heart. More willing to realize the light in all vibrations. Everything is part of us, arising out of the consciousness of the Creator. We can participate in the universal consciousness that contains and births everything, once we unlimit ourselves. All the difficulties we must face are designed to awaken us to our limitations, so that we can resolve them. Everything is a blessing, when understood from an expanded perspective. We can clear our consciousness and enter pure awareness of our presence of Being.

Our true Being has no physical presence. It is our present awareness, and is eternal and unlimited. We can express ourselves however we want in any dimension. Here we have expressed ourselves as our human selves, which are only imaginary creations of our true Self. We humans cannot know our true Being with our ego consciousness. We created the ego not to know, so that we could have a navigator in the realm of low vibrations and separation from our higher guidance.

We can direct our awareness beyond the ego by moving toward just being present in awareness. This is a state of no thoughts or feelings. Just awareness of everything without thought or emotion. In this state we can become aware of inner sounds, rhythms and the feeling of expansiveness. As we encounter people and situations, while we're in an elevated state of being, we intuitively know all about the energy confronting us. We can align with it, or invite it to align with us, if we are in a higher resonance. It is possible for any of us to be in a high-frequency etheric presence while participating in normal human life.

What happens now, as our limitations get resolved, is life begins transforming into a joyful experience. It doesn't matter where we are, or what our condition is, even the worst possible or the best, we're dealing with energy patterns, and we have the ability to modulate any energy patterns we encounter. We do this by putting ourselves into the situation that we want with our imagination and feelings. We can experience being in the scenario completely. This voluntary experience attracts energy patterns that are compatible with it, and life becomes better.

Living with Angels

As we live in increasing vibrations of energy coming from Gaia and the quantum field enveloping us, we are finding life becoming brighter and more beautiful. We are Being our true Selves more and more, and we are recognizing the higher expression of others. We can begin to recognize the consciousness of the Creator in every face. This is the only energy we need to interact with. Everything else has no inherent existence. It is synthetic, kept in existence with parasiticized human life force.

If we have any remaining belief in our personal conscious mortality, we can recognize it and resolve it through compassionate wisdom. We have the scientific proof of our timelessness. We need subconscious acceptance of our eternal nature, because it goes against our empirical experience. All of our personal challenges are designed to open our consciousness to timeless awareness.

How do we recognize the energy of the Creator in every face? In everyone we encounter in our experience or in our imagination, their energetic presence, their radiance, stimulates an emotion in us. This emotion has a frequency pattern of either low-vibrational fear or high-vibrational love. There is a neutral space between them, a space without fear or love. This is the perspective of our pure conscious awareness. From here we can instantly know the quality of any vibrations we choose to focus on.

We have absolute control over our choice of focus. By imagining that we are interacting with an angel, perhaps even disguised as an irritating person, we can imagine every word and motion is expressed by the Creator through this person. We can notice the frequency of the source that is creating this person and look into our own being for what aligns with that energy. This provides us with personal insight and a challenge for expansion.

We experience the energy of attraction. By choosing to imagine scenarios that we want to experience, and being in alignment emotionally, feeling ourselves in the situation we imagine as if it is actually happening, we are radiating that energy spectrum and attracting other energy patterns that bring us compatible experiences. To the extent that we can be absolutely present in every moment, our lives can improve proportionately. Once we fully embrace our eternal Being, there is nothing to fear in any dimension. We are free to love completely and be filled with joy, for we are expressing the vibrations of our Creator consciousness, and everyone we encounter can be angelic.

Energetic Alignments for Life Enhancement

For millennia our civilization has been primarily mind-based. Our emotions have been relegated to an untrained status in our personal relationships. Our egos have ruled the directions of our lives, and we have been disconnected from the energies of nature and from the Source of all life. The animals and the indigenous people have taught us that the Earth has energy patterns that we can know and align with. For instance, when there was a massive tidal wave a few years ago in Thailand, the elephants pulled up the stakes that held them captive and broke their bindings to escape with other animals into the hills well before the waves struck. The indigenous sea people tried to warn those on

Chapter 7. Aligning with Higher Consciousness

the beaches, but few listened. The energy of Gaia told them what was coming. Like the indigenous ones, we all have the innate ability to align with the Spirit of the Earth.

The spectrum of vibrations within the limiting boundaries of humanity's consciousness is held in expression and appearance by our group perception and recognition. Humanity's experiences are humanity's creation. We all have allowed or intended for our life force to be used to limit us to being slaves of fear. We have intentionally and unintentionally accepted fear of our mortality. In the face of our survival, we become desperate and either awaken or go through the mortality experience. Either way, we awaken to awareness of the energy patterns of Gaia.

To recognize Gaia's spectrum of energy, we can find natural places relatively undisturbed by humans. If we walk barefoot here, lie on the ground, put our ears to the earth, let the sun shine on us and the rain envelop us, listen to the birds and the wind in the trees, and just be here in open awareness, we can enter the energy frequencies of Gaia and intentionally resonate with her. Everything vibrating here arises out of unconditional love in the universal consciousness of the constantly creating One.

The natural energies for us are all in the frequency spectrum of love and joy. When we have no fear, because we believe and know that we are eternal Beings, we are free to live in higher vibrations. We can release every intention, except the intention to know the truth about everything. This quest leads to alignment with nature and our natural essence of Being. Our life force can become more powerful, we can be clearer, and we can become more acutely aware of our creative ability.

The Magic of Gratitude

Being thankful for everything can be life-enhancing in so many ways. Gratitude is a high-vibration state of being, and it attracts people and situations of compatible energy patterns into our lives. These we want to be thankful for. It is possible to think about being thankful, but unless we also feel it in our heart, it is only a vibration without intensity or amplitude. As in every situation in life, to be truly creative, we must have an interplay with our imagination and emotions.

Our situations may be bleak and difficult, but we still have the freedom to focus our attention on whatever we desire. Powerful emotions elicited by strongly challenging situations such as starvation or abusive relationships are difficult to transform in our awareness, but it is possible with strong intent. If we can isolate ourselves with inspiring music, hike to a waterfall pool for a swim or swim in the ocean, or do anything that may elicit inspiration, we can more easily change our focus.

We can be thankful for everything, including situations of intimidation and threats to our well-being. We can recognize that in the past we have intentionally and unintentionally created the vibratory pattern of every situation that we experience, and all serve a purpose for us. We are learning about ourselves and our creative ability. We're learning how to transform energy patterns into life-changing experiences.

We can be thankful for everything we are, all aspects of our personhood, our bodies, our environment, our friends and everything in nature. Gratitude is an easy feeling and intention to create. Because it is a high-vibrational state of being, it draws other high-vibration feelings and thoughts into our awareness. By living in this state of being, we can no longer feel compelled to focus on low-vibration situations and people.

Our personal energy signature vibrations attract into our experience people and situations that vibrate in resonance with us. By being in gratitude as much as possible and expressing our love and appreciation for everyone and everything, we create wonderful lives that we want to be thankful for. We can do this regardless of the chaos and violence that may be happening around us. We do not need to interact in alignment with low-vibration situations and people. Through a perspective of gratitude, we can transform all of our personal situations into experiences of peace, joy and abundance by following our inner guidance of what we feel best about.

Mastering Our Vibrations

In order to become masters of our lives, we must move beyond fear of every kind. The ultimate fear is termination of our conscious Being. If we can cure this misunderstanding of ourselves, we can eliminate all other fear.

Quantum physics has shown us that we interact with other energy patterns in the quantum field. Anything we imagine and recognize instantly appears for us on a subatomic level. If we can imagine that we are the energy of our subatomic particle/waves, then we experience the energy of our subatomic elements. We experience the realm beyond time and space, while our conscious attention is aware of timelines and their consequences. We can be aware of the true energy of our heart and know its guidance for us. We have the ability to imagine wonderful experiences and feel ourselves being in these experiences. If we have only these thoughts and feelings for a sustained period of time, we can learn to enter the reality of that spectrum of energy. It usually requires disciplined, intentional practice to make this leap in conscious expansion with confidence.

We know that we have energy signatures that manifest the bodies that we recognize. Energy is an eternal expression of universal consciousness. Energy can change vibration and form, but cannot be destroyed. Since we are the ones who create our personal energy signatures with the vibrations of our thoughts and emotions, we are also eternal, because what we create are expressions of ourselves. What we create is energy patterns that we experience.

The quality of our energy signatures can exist in the vibratory range of fear or the range of love. The dividing line between fear and true love is the consciousness of being eternal. It is the dimension of conscious awareness of the eternal now moment, which includes all moments in the quantum field. Here we can align with the energy of the Earth, whose frequency resonance is rising into new higher octaves. All humans who wish to continue to live on this planet must make the leap into Self-conscious eternal Being. The lower vibrations are now facing interference with the rising frequencies of the Earth and are weakening and dissolving without amplitude.

Actualizing Our Creative Impulse

Mystics and shamans have always known our creative genius, but Western scientists did not discover this ability until the rise of quantum physics, when it became clear that what we recognize appears out of the quantum field of infinite energy and potentiality. The actual creative mechanism appears to be inherent in our essential being and is a result of our nature as fractals, or exact replicas, of the universal Creator. We are everything that the Creator is, but not in our present form. As humans we exist in a very limited expression of our true Being, which we have access to in a deep meditative state of pure awareness.

Although many of us may not be aware of our larger Being, nevertheless we still exist as the creative essence of our essential Self. Every situation that we experience is a result of the quality of our thoughts and emotions, our hopes and fears, our visions and beliefs about ourselves and our destiny. By believing in fear of suffering and physical death, with the help of the media, we have created a pandemic. By believing in the primacy of love and eternal well-being, we can create the end of the pandemic. We do this not be resisting fear and repressive mandates, but by envisioning the lives we truly want and feeling ourselves living in loving and joyous situations and acting as if we are in the situations we envision.

Whenever we are in situations that are based in fear, we can realize that we have created these experiences for ourselves in order to learn how to use our creative abilities properly. Once we learn to use our thoughts and feelings with purpose and loving intent, we change our situations to manifestations of beauty, joy and abundance.

We are so close to being able to do this as a race, that we are on the verge of the awakening of all of humanity into a new world of our deepest desires and greatest love. We do not need to depend upon anyone or any circumstances outside of ourselves to do this. Our personal creative ability is a result of the inner light of the Creator flowing through our heart and enlivening our Being. Each of us is the Creator of the quality of our experiences, and we have the role of expressing this energy in every aspect of our lives.

Living Beyond Fear and Limitations

Fear is a construct of the ego, which has no higher guidance and feels separate and alone in attempting to survive in a competitive world. This is a world that we humans continuously create by our recognition of its energy patterns and alignment with its vibrations. Everything that exists, including the realm of human experience on this planet, is electromagnetic energy that is modulated by consciousness.

What we are capable of being aware of is determined by our beliefs about ourselves. Our beliefs are created by our egos as a result of our parental, social and religious training, indoctrination and personal preferences. They set the low-frequency boundaries of what we allow ourselves to experience. If we believe that we are biological entities with no connection to universal consciousness, we are in opposition to reality as quantum sciences have proved it to be, but our personal experiences will align with our beliefs. We will continue to create an unfulfilling and threatening world for ourselves.

All of our experiences here are self-created. Nothing exists unless it is held in the conscious awareness of a Being that arises from the essence of Creator consciousness, which is the source of everything. We have our own personhood and self-identity existing beyond time and space in many dimensions in eternity, and we project our consciousness into our physical world as the persons we believe that we are. It is possible for us to know our larger Selves by resolving our beliefs and allowing ourselves to become aware of a higher-vibration spectrum of energy.

We can understand that the experiences of fear, lack and suffering are self-created by our own thoughts and emotions according to our beliefs. When we focus primarily on confronting every situation with love, compassion, forgiveness and confidence in our own Being, we can change the energy around us into a higher dimension of expression. We can live in an environment of gratitude, beauty, joy and abundance just by holding these frequencies in our awareness and realization.

Our Innate Way of Knowing

We have lived by the guidance of our ego, the compartmentalized consciousness that we have created for ourselves in order to participate in the human experience. We have separated ourselves from our innate knowing. Since we are capable of creating an ego consciousness, along with our free will to use our abilities to create whatever we want, we have settled into a survival mode with each other, always needing something from others. We never have enough of everything we want to experience. We are the only species on this planet that is in this condition of lack. Our situation has to do with the use of our consciousness.

Chapter 7. Aligning with Higher Consciousness

The One Conscious Creator expresses Its Being through creatures like us. Just as a proton has the complete mass of the universe within its essence, according to quantum physicists, we also can know what is in universal consciousness, the complete consciousness of the Creator. A portion of this consciousness is the subconscious part of our being. It does not have free will and operates entirely in alignment with universal consciousness. Its abilities to manage our bodies are far beyond the capabilities of the ego.

Once we recognize the capability of our subconscious being, we have a clue as to the kind of capability our full consciousness must have. Realizing our capabilities requires our intentional recognition. We must align our thoughts and feelings with the vibratory frequencies of love, the essence of our Being, in order to be able to expand our awareness and abilities. Just as the birds, animals and fish have everything provided for them, including inner guidance for everything, so we were created to have everything provided for us, including inner guidance for everything. It is innate. We all have this inner guidance in our intuition. It is how we know anything, apart from what we can believe outside of ourselves. It is like the geese flying south for the winter. They all just know what to do in the moment.

We have this innate knowing. It needs our intentional recognition, which we can align with, when we imagine or experience high-vibrational, life-enhancing energy patterns. Out intuition is always present. We can become aware of it by loving our ego into submissiveness to higher guidance, which we can learn to pay attention to. Our intuition exists in the realm of unconditional love. Its vibrations are always life-enhancing. In every moment we receive inner knowing of how to interact with whatever energy we confront. There is always an inner light that naturally lives in the flow of unconditional love, joy, compassion, truth and beauty. We are moved by it always and can align completely with its energy.

Accelerating the Inner Journey

As the Shumann Resonance graph illustrates, the Earth is being engulfed in massive waves of gamma ray photons from our Sun and from our surrounding cosmos. These waves are raising the frequency of our planet and all conscious beings living here. As free-will Beings, we have the option of choosing what frequency we want to focus on. From our experience we know all about focusing on fear-based feelings. It's time to change focus. Gaia, the Being manifesting itself as our planet, is raising her frequency into a world without fear. We are all flowing into love.

By understanding the significance of the Earth's resonance, we can know that the energy patterns of feeling grateful and joyous is growing stronger. Everything is getting brighter, to the point that there is an increasing trend for unveiling the dark ones. Once they are widely recognized, they will depart, because their energetic bodies cannot viably be subjected to powerful high-vibration resonance. Vibrationally they must come into alignment with the higher vibrations or become unstable and dissolve. It is not our destiny to continue in a low-vibration environment. Everything must rise in vibrations into the world of only love.

Analyzing the trends of conscious changes, we may conclude that the low-vibration world of human experience with fear is disappearing, due to our loss of fascination with it. We can change the focus of our attention to forgiveness, gratitude and compassion in every moment and in every encounter. This is the way of love and fulfillment.

Every energetic resonance pattern in our experience has our life force keeping it in its pattern of vibration as a result of our focus of attention to it. We can dissolve or transform every low-frequency situation that confronts us by changing our energetic focus, which withdraws our life force from maintaining any life-taking energy patterns. By refocusing on high-frequency situations, we can raise our own energy signatures and attract high-vibration experiences. The vibration of our focus of attention determines the quality of experience we are creating in each moment of our awareness.

We can clear the quality of our vibrational focus of attention by entertaining in our constant present awareness only life-enhancing thoughts, imaginings and feelings. This creates a leap in consciousness into a higher dimension of life.

Discerning Between Ego Mind and Intuition

Upon incarnation we begin developing our ego consciousness. We also arrive with our intuitive knowing. These are not the same, and many of us learn to rely upon our mental processes of ego consciousness to make our way through human life. Unlike our ego mind, our intuition does not intrude on our awareness. It requires our recognition.

We develop our ego consciousness as a result of our feeling of separation from the Source of our Being. We feel alone and vulnerable. We become fearful, not really knowing what challenges will confront us or how to react successfully to all of the threats. We form beliefs about our limited abilities and the duration of our lives. We resign ourselves to believing in our ultimate demise. In our ego mind we inhabit a spectrum of low-frequency energy that in its essence is destructive to life. This is what we came to this planet to experience and to find out what it feels like, so that we can develop deeper compassion.

If we decide to take a few deep breaths and step back mentally and emotionally, so that we can observe ourselves in our situations, it becomes possible for us to search for a deeper meaning in life. In our innermost Being we already know everything about life. This knowing is our intuition. It is our essential Being beyond time and space. It is part of our multidimensionality, which the ego mind cannot grasp, but which we can realize.

When we allow ourselves to resolve our fears and limiting beliefs, we can open ourselves up to higher vibrations. We can feel love and joy emanating from the heart of our Being, symbolized by the energy of our physical heart, which lives only to enliven us and enhance our lives. This is our clue to recognize our intuitive knowing. It is high-vibration life-enhancing energy that feels wonderful and confident in our eternal Being.

In this life we need both our ego and our intuition. Our ego is always present and must be trained to allow our intuitive awareness. In recognizing and following our intuitive knowing, our lives become inspired and open to our essential Being as aspects the universal consciousness of the Creator of all.

On Being Who We Really Are

Our entire cosmos has an energy signature. Every galaxy has its own energy signature, and so does every sun and planet. All of them are conscious beings, just like us, but with different manifestations created in their essential Being, just like us. All of them arise from the same universal consciousness that we do. Gaia, the Spirit of our planet, has her own energy signature, which is partially displayed on the Shumann Resonance Graph. This graph shows us the energy patterns that will be manifesting in our empirical world shortly. As the vibrations change, the physical manifestations change. The quality of these changes indicates the changing vibrational quality of the Earth. This is the energy spectrum that we live within. If we are out of resonance with it, we feel as if we have no heart-felt connection with Gaia. If we can be people of the land, even occasionally, and walk barefoot on the earth, swim beneath waterfalls, hike through forests, dive into clear waters and climb high mountains, we can align with what we feel in these immersions in nature. We intuitively can align with Gaia's vibrations, if we are open to feeling them. Many indigenous people have this connection, but most "civilized" people do not. We can recover it with our intentional openness to the conscious presence of Gaia radiating from her planetary body.

Humanity has its energy signature. All of the different races and individuals contribute to its vibrational spectrum. The higher vibrations are the more powerful, but we designed the human experience to occur predominantly in low-vibratory frequencies of various forms of fear. Now humanity is out of alignment with the rising frequencies of Gaia. As a result, there is instability, chaos, political insanity and great oppression, while we process the adjustment to higher frequencies. Those who choose to stay in the traditional spectrum of humanity's energy signature decide to continue life as they know it. If we choose to live high-vibrational lives, we can do so. We must make the choice and recognize high-vibrational experiences. Life becomes filled with enjoyment, compassion, peace, gratitude, love and abundance in every way. This all manifests just by our way of Being. We live in these emotional states and attune to the energy of the heart of our Being in our thoughts and feelings. We do this in every encounter we have and every kind of energy that we face. In this way we transform ourselves into the One we were created to be, and we can trust ourselves to use our infinite creative power wisely and in energetic alignment with high frequency scenarios.

Cultural Identity and Consciousness

Indigenous cultures use oral teachings and story-telling that stimulates inner realization to connect with the energies of the Earth and our telepathic intuitive knowing. In an environment that is in balance with the energies of the Spirit of the Earth, and where nature spirits are recognized, people of the Earth know how to live and survive harmoniously in nature and are often guided by dreams and visions. The shamans have even expanded their awareness beyond the physical body. There is still, however, limited cultural awareness of the magnitude of our expansive Being. Every culture has developed limiting beliefs that keep us from knowing our real potential, and as a result we do not penetrate the deepest knowing of our abilities and our essence.

In cultures that depend upon the written word and technological communication, much more knowledge is available to our conscious minds, but the connection with our intuitive knowing has been weakened. We have learned that we acquire knowledge from outside of our own Being. We have been trained to believe that mental prowess is our most important quality for living successfully. We have diminished our connection with Gaia and no longer recognize the spirits of nature. This has resulted in less love in our lives and disorientation with the meaning of life, because we don't understand how life works and how it happens.

We can come from any culture and become aware of our unlimited creative abilities by understanding that we express ourselves as patterns of electromagnetic energy that our consciousness interprets as material experience. We can consciously work with energies. We constantly create energetic patterns with our thoughts and emotions. We are modulators of energy. We determine the frequency and amplitude of the patterns that we envision and feel in every moment. These manifest for us as the qualities and intensities of our experiences.

By developing a perspective of always being in high-frequency vibrations of emotions and visions, we begin to live in experiences that feel thankful, compassionate and loving. Our habitual thoughts and feelings create an energy signature that interacts with other energy signatures in the quantum field and attracts those with compatibly-aligning frequencies. These become our experiences. We are not limited in our Being to any spectrum of frequencies in our abilities to think and feel. As we intentionally open ourselves to higher frequencies in the spectrum of gratitude, love, abundance and joy, we begin to attract those energies which become our experiences.

The Emergence of Our Inner Power

Whatever we experience in our lives is what we have created by the radiance of our personal energy signature. Life appears to happen on an empirical level, but really the physical world is only our human consciousness interpreting the qualities of vibrations of the energy patterns that we encounter in the quantum field of all potentialities. The energy that we encounter is attracted to us by our own vibratory patterns, which are determined by our thoughts and emotions and limited by our beliefs.

We are the masters of our lives. We are constantly creating the qualities of our experiences with the focus of our attention and the projection of our emotions. We are enveloped in a plasma energy field that constantly interacts with us. We can't see it or feel it, but its invisible presence brings into manifestation in our lives everything that we experience through electrical and magnetic polarities.

Our physical bodies operate by electrical impulses stimulated by our conscious and subconscious thoughts. Our emotional bodies operate by magnetic impulses emitted by how we feel in every moment. Our thoughts and feelings are uniquely ours. Nothing outside of ourselves can control them. They are aspects of our free will. They can be elicited by our physical experiences, but we are not compelled to respond in any specific way. We are completely free in our own Being.

Quantum physics has shown us that all conscious beings constantly arise beyond space and time from the essence of the universal consciousness that expresses itself as the quantum field of all possible energy patterns that can manifest in our experience. We participate in this consciousness and have in our own true essence all of the abilities inherent in universal consciousness.

In our essential Being, we are unlimited in every way, and we can awaken from our compartmentalized conscious awareness in the world of our current human experience to the unlimited world of universal consciousness that holds our expanded personal identity.

By resolving our self-imposed limitations that we have assumed in order to have a human experience on this planet, we can open ourselves to expanded conscious awareness. In this way we can understand everything that happens in our lives as a result of our own mental and emotional attractions. If we feel restricted in any way by governments or any authorities outside of our own Being, their mandates exist for us only as stimulants for us to awaken to our own true essence. By maintaining a high-vibration perspective of compassionate understanding, gratitude and love, we can resolve all low-vibrating energies that we encounter. It is time for us to stand in our innate personal power, which cannot be challenged. We can create whatever vibrations we need and desire to manifest in our experience.

www.ingramcontent.com/pod-product-compliance
Lightning Source LLC
Chambersburg PA
CBHW032039150426
43194CB00006B/349